SpringerBriefs in Materials

The SpringerBriefs Series in Materials presents highly relevant, concise monographs on a wide range of topics covering fundamental advances and new applications in the field. Areas of interest include topical information on innovative, structural and functional materials and composites as well as fundamental principles, physical properties, materials theory and design. SpringerBriefs present succinct summaries of cutting-edge research and practical applications across a wide spectrum of fields. Featuring compact volumes of 50 to 125 pages, the series covers a range of content from professional to academic. Typical topics might include

- A timely report of state-of-the art analytical techniques
- A bridge between new research results, as published in journal articles, and a contextual literature review
- A snapshot of a hot or emerging topic
- An in-depth case study or clinical example
- A presentation of core concepts that students must understand in order to make independent contributions

Briefs are characterized by fast, global electronic dissemination, standard publishing contracts, standardized manuscript preparation and formatting guidelines, and expedited production schedules.

More information about this series at http://www.springer.com/series/10111

Nikhil Padhye

Molecular Mobility in Deforming Polymer Glasses

Theories and Applications

 Springer

Nikhil Padhye
Stoneham, MA, USA

ISSN 2192-1091 ISSN 2192-1105 (electronic)
SpringerBriefs in Materials
ISBN 978-3-030-82558-4 ISBN 978-3-030-82559-1 (eBook)
https://doi.org/10.1007/978-3-030-82559-1

This Springer imprint is published by the registered company Springer Nature Switzerland AG
The registered company address is: Gewerbestrasse 11, 6330 Cham, Switzerland

To David M. Parks on his recent 70th birthday.

Preface

Plastic deformation in glassy polymers and chain mobility in polymer melts are usually studied under two fields of specialization: continuum plasticity and polymer physics. However, the molecular motion associated with plastic deformation in polymer glasses is a key ingredient to gain a fundamental understanding of the mechanical behavior of polymers at macroscopic and microscopic levels. This monograph attempts to bridge the gap between these two scales by undertaking a unified outlook at an introductory level, provides the necessary background in polymer physics and polymer mechanics, and discusses the topics of molecular mobility accompanying macroscopic inelastic deformation in glassy polymers.

Care is taken so that the content caters to an audience with diverse backgrounds. This monograph focuses primarily on the class of glassy polymeric materials; however, references to other materials are provided only if relevant or necessary. Despite some qualitative similarities with glassy polymers with regard to local microscopic structural rearrangements of atomic clusters during inelastic relaxation, I do not pursue the discussions on metallic glasses because they generally exhibit much less ductility. The topics of material mechanics and enhanced polymer mobility in case of semi-crystalline polymers are also not taken up primarily because there are no "smoking gun" direct applications of deformation-induced mobility in this class of materials. In the solid state, the plastic deformation in semi-crystalline polymers requires activation of slip systems within crystalline domains, or even twinning or martensitic transformations in the presence of surrounding amorphous interphases, which makes the issue of deformation-induced mobility more intricate. At high degrees of crystallinity, semi-crystalline polymers are brittle. Therefore, without pronounced ductility, the deformation-induced mobility will be severely restricted. Moreover, blending of crystalline polymers with additives to enhance deformation-induced mobility for bonding or other applications (as discussed in this monograph) will likely disrupt the internal crystalline structure and render a behavior similar to that of amorphous materials. Notwithstanding the current state of affairs, deformation-induced mobility and its applications in semi-crystalline polymers remain to be investigated.

This monograph is expected to facilitate an introduction to academic coursework and research on the physics, mechanics, and plasticity in the context of glassy polymers. It is also well-suited as supplementary reading in a graduate-level course on the mechanical behavior of polymers. Although the treatment is kept quite brief, I have provided several elementary concepts that unify different fields so that readers from different disciplines can understand this topic, and researchers, students, and practitioners can apply the concepts of the mechanics, plasticity, and diffusion of polymers to various engineering applications. I foresee the central topic of this monograph, deformation-induced bonding, as an upcoming and independent field, with promising areas of research and applications in modeling and simulations at the molecular and continuum levels, and advanced analytical characterization. Open questions in this field of research are discussed and emphasized throughout the text.

Due to the complexity of the molecular-level processes associated with the continuum-scale plastic deformation in polymers, much to my disappointment, at this stage there are no exact, deterministic, and provably correct experimental means or molecular-level theories that precisely reveal all molecular-level events and quantitative characteristics of motions that occur during large strain plasticity in polymers. Nuclear magnetic resonance (NMR) has emerged as a powerful tool in revealing certain dynamical relaxation characteristics in polymers. Other techniques, such as small-angle X-ray scattering (SAXS), wide-angle X-ray scattering (WAXS), Raman spectroscopy, and transmission electron microscopy (TEM), have been used judiciously to study the microstructure (in situ or post-deformation) for inferring structural changes. However, complete molecular-level understanding of deformation processes is far from complete. With advances in high-performance computing, atomistic and molecular simulations have started having a crucial role in providing plausible molecular descriptions associated with the observed macroscopic behavior. Limitations on accessible timescales and length scales made feasible through simulations (even with the most powerful computing platforms), and the unavailability of exact and direct experimental means for exhaustively monitoring molecular-level activity for each polymer molecule with complete precision in an arbitrary solid-state glassy deformation, are major challenges. We caution that, in general, mechanistic inferences drawn from molecular-scale simulations or experiments of glassy polymers should be considered as suggestive, and care should be taken before drawing exacting conclusions for various reasons.

The underlying challenge in any molecular modeling associated with the plasticity of glassy polymers stems from the fact that a polymer glass is in a non-equilibrium state. Even prior to deformation, the state of the polymer molecules is not certain (i.e., it is dependent on the processing history, and the subsequent deformation states themselves are not in a thermodynamic equilibrium). This characteristic of polymer glasses is in contrast to polymer melts or polymer solutions, whereby modeling of molecular motion in a thermodynamic equilibrium (with appropriate assumptions and approximations) has been tractable and shown some remarkable successes (e.g., for predicting scaling laws for dependence of viscosity or diffusion coefficients on the molecular weight). Unless direct and definitive means of probing molecular-level activities are established and fully verified, deciphering precise molecular information during plastic deformation will remain unsolved. In spite of this predicament, molecular models, theories, experimental techniques, and simulations that can explain observable findings are useful from a conceptual perspective.

With this background, I have found no single monograph that focuses on deformation-induced molecular mobility in polymer glasses in a detailed and consistent manner while covering and linking the relevant, elementary, and inter-disciplinary ideas. There are articles and book chapters available on this topic, but they are not self-contained or tutorial in nature. Most of the excellent textbooks on the mechanical behavior of polymers do not focus on the significance and mecha-nisms underlying deformation-induced mobility. What is also missing is a holistic viewpoint that connects critical concepts in polymer physics, polymer mechanics, continuum plasticity, and molecular mobility in a simplified (yet complete) manner. It is with this goal in mind that this short project was undertaken. However, due to the limited scope of this monograph, references and discussions on several important works in the physics, mechanics, and material modeling of polymers have been undesirably omitted. Similarly, detailed modeling of the mechanical behavior of polymers, continuum mechanical frameworks for large-strain plasticity, numerical algorithms, and extensively available test data have not been discussed. It is expected that, after reading this monograph, readers will be able to study related topics in detail by following pointers to external references.

This monograph is based on the research work that started at the Massachusetts Institute of Technology (MIT, Cambridge, MA, USA). Its success was truly a result of superb technical guidance provided by Professor David M. Parks. David deserves the credit for identifying and defining the phenomenon of sub-glass transition temperature (T_g), solid-state, plasticity-induced bonding based on the experimental results submitted by me as part of a term-project report in a graduate class at MIT. David is acknowledged for educating me on the plasticity and mechanics of polymers, for which (and many other things) I am truly grateful! David's authority, combined with simplicity, elegance, and clarity, has been a continual source of inspiration for this and related work(s). His unassuming, kind, and warm persona can only be overshadowed by unfathomable depths of his true knowledge and unmatched natural brilliance. This monograph has been assembled in the spirits to celebrate David's recent 70th birthday. It is a tiny tribute by a former student to his teacher. It is expected that the insights offered through this monograph will aid wide applications involving polymer adhesion (explicitly or implicitly). I credit that success to David, with whom an interesting and enlightening discourse has continued on these (and related) matters.

Contributions by the late Professor Ali S. Argon and colleagues at MIT on the mechanics and continuum plasticity of polymers (which essentially laid the foundation for the modern continuum computational plasticity of polymers), and in part for this work, are acknowledged. Professor Lallit Anand's coursework and study material on continuum plasticity at MIT is acknowledged: it provided a formal exposition of this field to me in my formative years. Beginning recently, discussions with Professor Gregory C. Rutledge from the Chemical Engineering

Department at MIT on molecular simulations are appreciated. Some of those discussions and exchanges made their way into this monograph and shaped it, and are acknowledged.

Nikhil Padhye

Contents

Chapter 1
Polymer Physics and Dynamics of Polymer Melts

Abstract The random motion of a flexible molecular chain floating in a solvent or a melt can itself be quite complex. The discussion in this chapter starts with some basic concepts of an ideal polymer chain. It is then followed by classical models describing the dynamics and diffusion of polymer chains applicable to polymer melts. This chapter's main objective is to contrast the differences between the mobility of polymer chains in melts (or solvents) and the kinetically trapped glassy state.

1.1 Introduction

Polymers are commonly referred to as macromolecules because they are large molecules comprising smaller repeating units (monomers) which are bonded covalently. Depending on the arrangement of monomers, the polymers can be categorized as linear, branched, or crosslinked. In linear polymers, the monomer units are repeated along the backbone of a polymer chain. Branched polymers have branches of molecules that are covalently bonded to the main chain. In crosslinked polymers, the molecular entities of one polymer chain are covalently bonded to that of another polymer chain, thereby giving it a network structure. Elastomers contain loosely crosslinked networks, whereas thermosets have a high degree of crosslinked networks. Thermosets degrade upon heating, whereas thermoplastics (which do not contain crosslinks) melt upon application of heat. For other types of classifications in polymers see [29]. The molecular architecture of a polymer directly controls its physical and mechanical properties, so special attention must be given to it for describing polymer properties. The topics discussed in this monograph are focused primarily on linear polymers.

The molecular-level activity of polymer chains is dependent (among other things) upon the physical state of the polymer (i.e., whether the polymer is in the solid state or in a melt or solution). The concepts of solid-state glasses are discussed in greater details in the next chapters. However, let us start with discussion of the dynamics of a single isolated polymer chain and use it as a "building block" for more sophisticated theories that consider the interaction of one polymer chain with other macromolecules.

1.2   Dynamics of an Ideal Polymer Chain

One of the simplest treatments of a polymer chain is to idealize it as a completely flexible (freely jointed) chain. The "ideal" chain is also known as the Gaussian chain. Figure 1.1 shows an ideal linear polymer chain comprising several monomers. The start and end points are marked by α and ω, respectively. The chain comprises N monomers, each of length a units. From the statistics of a random walk [6], it can be shown that the mean square end-to-end vector $\langle r^2 \rangle = Na^2$, or $\langle r^2 \rangle^{1/2} = \sqrt{N}a$. If α is considered to be origin, and R is the end-to-end distance, then the corresponding probability distribution function is given as

$$P(N, R) = \left(\frac{3}{2\pi Na^2} \right)^{2/3} exp \left(\frac{-3R^2}{2Na^2} \right). \tag{1.1}$$

Accordingly, the conformational entropy of this ideal chain is given as

$$S(N, R) = S_o - \frac{3K_B R^2}{2Na^2}, \tag{1.2}$$

and the free energy of the chain is given as

$$F(N, R) = F_o + \frac{3K_B T R^2}{2Na^2}. \tag{1.3}$$

In Eqs. 1.2 and 1.3, S_o and F_o are entropy and free energy, respectively, in some reference configuration. Major aspects of the motion of a flexible polymer chain can be understood in terms of a "random walk on a periodic lattice", as illustrated schematically in Fig. 1.2. The detailed derivations can be found in several popular references, for example, [6].

Fig. 1.1 Polymer chain

Fig. 1.2 Random motion of a polymer chain

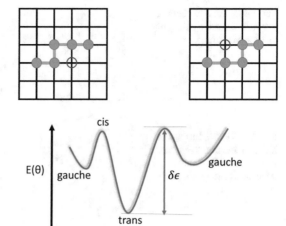

Fig. 1.3 Energy barriers for different polymer conformations

In reality, the rotation along the polymer backbone requires overcoming an internal barrier because the chains are not fully flexible, as illustrated in Fig. 1.3. Due to these barriers, the flexibility of the polymer chain is inhibited. However, at high temperatures, the assumption of ideal flexibility holds fairly well. According to the Boltzmann distribution law, if the energy associated with some state of a system is ϵ, then the frequency of occurrence of that state is proportional to $e^{-\epsilon/K_B T}$. Thus, if $K_B T$ is large, then all possible energy states are equally likely. This situation implies that at high temperatures, the transition from *cis* to *trans* or vice versa, or of any transition from one conformation to the other, can be achieved readily. If thermal activation is sufficiently high compared with the activation-energy barrier between different energy states, then all possible conformations are equally likely; this is also referred to as "dynamic flexibility". This flexibility cannot be obtained in non-deforming solid-state glass.

For successive C-C links, if the transition from *trans* to *gauche* requires a characteristic time τ_p, then $\tau_p = \tau_0 exp(\frac{\delta\epsilon}{K_B T})$, where $\delta\epsilon$ is the activation energy of the transition from trans to gauche, as shown in Fig. 1.3. The reference time scale τ_0 is approximately on the order of 10^{-11} s (timescale for atomic vibration). If a polymer chain has dynamic flexibility, then transitions from one conformation to another occur very rapidly. Lastly, according to the concept of free volume (discussed in greater detail in the next chapter), if the temperature increases then free volume also increases, due to which transitions from one conformation to another are aided.

An important concept is that of the persistence length (L_p). Persistence length is a characteristic length for loss of orientational correlations along the chain. In other words, L_p is a characteristic length beyond which one can see the polymer chain as continuous and flexible. If $L_p << L$, where L is the length of the polymer, then the molecule is quite flexible at large scales. If $L_p >> L$, then the polymer is not flexible. This scenario also leads to the concept of "static

Fig. 1.4 Interaction between polymer segments belonging to a single chain

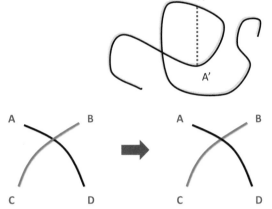

Fig. 1.5 Real chains cannot intersect and crossover each other, a consideration not taken into account for ideal chains

flexibility", which indicates the scale at which a polymer chain can be viewed as a random coil. Importantly, one can find polymer molecules which are flexible from a static viewpoint. That is, if observed at larger scales they appear to be a coil but, if the activation-energy barriers for transitions are high, then the situation corresponds to a random coil, which is essentially frozen in one conformation. In the glassy state, polymer chains are essentially "locked" in their conformations, and the activation energies for the conformational transitions are quite high compared with those for thermal activation, so a chain will appear rigid (albeit for atomic-level vibrations). The assumption of an ideal chain also ignores any monomer-to-monomer interactions. As shown in Fig. 1.4, the interactions between AA' segments are neglected for an ideal chain. However, for a real chain such effects are present and were first recognized by Flory as "excluded volume" [13]. These are also referred to as "volume interactions".

Due to excluded-volume effects, a real polymer chain occupies more space than that in an ideal chain conformation. Another way to describe this would be to turn the inter-atomic potential "on" and "off" on an ideal Gaussian coil. The concept of excluded volume implies that there is an interaction between monomers, and so the net volume occupied by a chain is different from an that of an idealized chain. This is also related to the idea of "self-avoiding walks" (SAW). That is, in reality, polymer chains (or segments) cannot intersect with themselves. This is shown schematically in Fig. 1.5.

An ideal-chain treatment predicts $R \approx N^{0.5}$ but, for real chains, $R \approx N^{3/5}$, i.e., a different scaling is noted [6]. However, in the solid state or in polymer melts, the excluded-volume effects are screened (i.e., negated) and then the scaling reverts to $R \approx N^{0.5}$. Lastly, the status of chain configuration in a solvent depends on the nature of the interaction between the polymer and solvent. Dilute polymer chains in a good solvent swell such that $R \approx N^{3/5}$, rather than $N^{1/2}$.

1.3 Dynamics in Polymer Melts and Self-Diffusion

The diffusion coefficient is a quantitative measure for the mobility of polymer chains. Broadly speaking, the diffusion coefficient can be described for various polymer systems (e.g., polymeric melts, polymeric liquids, dilute polymer solutions) and the diffusion of species into polymers. The molecular weight, polymer concentration, physical state, and temperature affect the diffusion coefficient. In this monograph, we are interested primarily in self-diffusion, i.e., diffusion of system A into A itself, where A represents polymer molecules of a particular type. In subsequent sections, we shall focus mainly on self-diffusion in the context of polymer melts (or solutions) and models describing them.

Polymer melts (or solutions) are an equilibrium system. Their mobility is commonly described by the models of Rouse [24] and de Gennes [5]. Zimm's model [30] is an improvement over the Rouse model because it takes hydrodynamic interaction into account. A detailed review of the tube models of diffusion in polymers is given in [21].

The fundamental dynamic property that characterizes the average motion of a polymer chain is the self-diffusion coefficient (D). The tracer diffusion coefficient (D^*) denotes a situation in which the background, in which the motion of a tracer particle occurs, is made up of neighbors different than the tracer particle. The diffusion coefficient is sensitive to molecular-structure variables such as components, molecular weight, and branching, and it estimates the time-averaged displacement of the center of mass of the molecule. In three-dimensional (3D) space, a diffusive motion is described as $\langle |\mathbf{r(t)} - \mathbf{r(0)}|^2 \rangle = 6Dt$, where $\mathbf{r(t)}$ and $\mathbf{r(0)}$ denote the position vectors of a diffusive particle at a time t and the onset, respectively. In one-dimensional (1D) motion, if x denotes the distance coordinate, then the diffusive motion is given as $x^2 = 2Dt$. The diffusion coefficient depends directly on the temperature.

Next, we discuss the basic concepts and ideas underlying the Rouse model and reptation model of diffusion. They remain the most popular models to describe polymer diffusion in melts or solutions, and help to contrast the motionless state in solid-state glasses. This strategy will enable better understanding of the notion of deformation-induced polymer mobility and bonding.

1.3.1 Rouse Model

The Rouse model [24] is derived from a description of springs and beads suspended in an isothermal temperature bath (Fig. 1.6). The potential energy and kinetic energy of this bead-spring model is fixed, and the springs and beads do not interact with each other while moving dynamically. In the Rouse regime, polymer chains are assumed to be unentangled, and the diffusion coefficient $D_{rouse} \sim N^{-1}$, where N is the number of monomers. The Rouse model is considered to be the first molecular model for polymer dynamics.

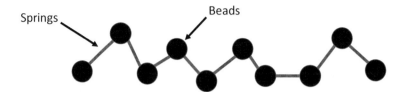

Fig. 1.6 Rouse model based on springs and beads

We derive this relationship by scaling analysis. First, we calculate the time scale τ required for a polymer coil to move a distance on the order of its own size R using scaling arguments. Under a simple diffusive motion,

$$\tau \approx \frac{R^2}{D}. \tag{1.4}$$

According to the Stokes-Einstein equation, the diffusion coefficient is given as

$$D = \frac{K_B T}{\chi_o}, \tag{1.5}$$

where χ_o denotes the drag coefficient on the entire polymer coil and K_B is the Boltzman's constant. By substituting 1.5 in 1.4 we get

$$\tau \approx \frac{R^2 \chi_o}{K_B T}. \tag{1.6}$$

If we assume that the total frictional drag on the chain is the sum total of frictional drag on N segments, then $\chi_o = N\chi$, where χ is the friction drag coefficient on one segment. Thus,

$$\tau = \frac{R^2}{D} = \frac{R^2}{\frac{K_B T}{N\chi}} = \frac{\chi N R^2}{K_B T}. \tag{1.7}$$

For a freely jointed chain we know that $R^2 = Na^2$, therefore substituting R^2 in Eq. 1.7, we get

$$\tau = \frac{\chi N^2 a^2}{K_B T} \tag{1.8}$$

or,

$$\tau = \tau_0 N^2. \tag{1.9}$$

Here, τ_0 denotes the time scale for motion of individual monomers. That is, τ_0 is the time scale at which the monomer would diffuse a distance of order of its size (a) if it were not attached to the chain. Now we estimate τ_0,

$$\tau_0 \approx \frac{R_{monomer}^2}{D_{monomer}} \tag{1.10}$$

and,

$$D_{monomer} = \frac{K_B T}{\chi} \tag{1.11}$$

where, $R_{monomer}$ indicates the radius (or the size) of the monomer (which is a). If a monomer is in a solvent with viscosity η_s, then the monomer friction coefficient based on Stokes law can be used to estimate χ, i.e., for the monomer

$$\chi \approx 6\pi a \eta_s. \tag{1.12}$$

Thus, by substituting 1.12 in 1.11

$$D_{monomer} \approx \frac{K_B T}{\eta_s a}. \tag{1.13}$$

Therefore,

$$\tau_0 \approx \frac{\eta_s a^3}{K_B T}. \tag{1.14}$$

Hence the Rouse relaxation time (τ_R) is given as

$$\tau_R \approx \frac{\eta_s a^3 N^2}{K_B T}. \tag{1.15}$$

This also leads to the scaling law for the diffusion coefficient based on the Rouse model as

$$D_{Rouse} \approx \frac{R^2}{\tau_R} \sim \frac{N}{N^2} = \frac{1}{N}. \tag{1.16}$$

The treatment described above considers the motion of the polymer coil without paying attention to its internal relaxations. To solve a full relaxation spectrum, the dynamics of a Rouse model with the chain as a succession of "beads" $r_1 \ldots r_n \ldots r_{n+1}$ separated by "springs" along the vectors $a_1 \ldots a_N$, must be considered, as shown in Fig. 1.8. The chain is assumed to be long enough to obey Gaussian statistics, and the entropic energy of the chain is represented by the springs. This idealization is shown in Fig. 1.7. This system of beads connected with

Fig. 1.7 Representation of
flexible, randomly coiled,
macromolecules using the
bead-spring model

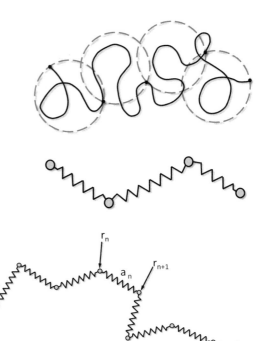

Fig. 1.8 Spring and bead
model describing Rouse
motion

springs will have a solution in terms of eigen-modes, which describes the relaxation
of an elongated state of a chain. A detailed derivation and discussion is given in a
standard textbook [6]. The system shown has a set of relaxation times τ_p (p indicates
the mode of relaxation), that scales as (Fig. 1.8)

$$\frac{1}{\tau_p} = 3\pi^2 \frac{T\mu}{a^2} \left(\frac{p}{N}\right)^2 = W \left(\frac{p}{N}\right)^2. \qquad (1.17)$$

Equation 1.17 is a quadratic dispersion relationship ($1/\tau_p \sim p^2$). Here, μ is
the "mobility" (the ratio of the monomer's terminal drift velocity to an applied
force). Note that the longest relaxation time ($p = 1$) scales like N^2 in the Rouse
model. Importantly, on timescales shorter than the Rouse time, the chain exhibits
viscoelastic modes. However, on timescales longer than the Rouse time, the motion
of the chain is simply diffusive. Such diffusive motion of the entire chain is severely
restricted on experimental time-scales in a solid-state glass.

The different modes of relaxations should be clarified. If we observe the motion
of polymer for times greater than τ_R then it would be seen to be diffusive, i.e., if the
observation window is greater than τ_R then, according to Rouse model the motion of
the center of mass of the polymer coil would be given as $\langle r_{cm}^2 \rangle = 2D_R t$, where D_R is
the diffusion coefficient according to the Rouse model. When probing on timescales
less than τ_0, the polymer segments essentially exhibit an elastic response because

Fig. 1.9 Stress relaxation
and Rouse relaxation time

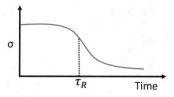

all modes of cooperative relaxations have timescales that are greater than τ_0. On
timescales greater than τ_R, the polymer moves diffusively and exhibits the response
of a simple liquid. An ideal viscoelastic response of any polymeric liquid should
capture the full-spectrum of relaxation times (and may require detailed modeling).
If we apply a step strain on the polymer, then each monomer stores an energy $\approx$
$K_B T$. If the whole chain relaxes (i.e., releases the stress by reorganizing, as a whole,
on the macro-molecule scale), it qualifies as the Rouse relaxation. This is shown in
Fig. 1.9.

Polymer dynamics in the melt state (with no solvent) is described by the Rouse
model for short chains that are not entangled. The hydrodynamic interactions in
polymer melts are screened (just as excluded-volume interactions are screened in
melts). The other limit of the unentangled polymer dynamics is the Zimm limit,
which applies to dilute solutions where the solvent within the pervaded volume of a
polymer is hydrodynamically coupled to the polymer. Researchers have reported a
deviation in the scaling law of $D \sim 1/N$ for unentangled systems based on Rouse
prediction when free-volume effects occur.

1.3.2   Reptation Model

The Rouse's model (1953) has several limitations because it assumes an ideal
behavior of the polymer chains, and locality of the response. The chains in the Rouse
model can crossover each other, and therefore cannot describe the behavior of the
entangled melts or solutions. De Gennes was interested in describing the motion
of polymer coils trapped in a network, and proposed a model of reptation (1971). A
series of tube-models for the rheology and diffusion of polymers were also proposed
by Doi and Edwards [31].

As an illustration, Fig. 1.10 shows a chain 'P' moving in a given network (e.g.,
rubber or elastomer). Here, O_1 and O_2 are fixed obstacles that the chain cannot
cross. The motion of the chain 'P' is worm-like, and is called "reptation". The
idea of reptation has also been extended to describe diffusion in polymer melts by
replacing the concept of network (rubber or elastomer) with a melt of chains that
are chemically identical but very long in length such that even under stretch the
long chains in the melt do not disentangle.

In the reptation regime, polymer chains experience topological constraints, and
lateral displacements in the entangled chain structure are negligible compared

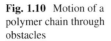

Fig. 1.10 Motion of a polymer chain through obstacles

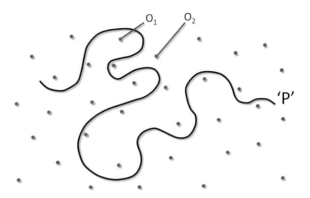

with longitudinal diffusion or "reptation" in the "tube" formed by neighboring molecules, with $D_{reptation} \sim N^{-2}$ (or $D_{reptation} \sim 1/M^2$, where M is the molecular weight of the polymer). We emphasize that the chain is confined to a tube while having a shape similar to a random-coil configuration of the chain. The only thermodynamically favorable motions are along the chain-contours. For polymer melts, the transition from the Rouse regime to the reptation regime is often noted at the critical entanglement length of the polymer chain. For an illustration, see [16]. Evidence of reptation through experimentation is provided in [18]. Measurement of the self-diffusion coefficient of polystyrene (PS) chains in benzene using forced Rayleigh scattering was carried out in [15]. Here, the semi-dilute regime chains overlapped, and the diffusion coefficient showed a compatible behavior with the scaling-law predicted by reptation.

The constraints forming a tube will themselves reptate during the residence time of the diffusing chain in this tube. This mechanism, referred to as "constraint release", was identified by Graessley [14], and enables one to account for the molecular mobility of the chain of interest by allowing the topological environment to vary due to reptating surroundings. An experimental study on reptation and constraint release in linear polymer melts was made in [27]. The authors showed that reptation was the dominant mechanism above a molecular weight of $10M_e$ (M_e is the entanglement molecular weight), whereas below $10M_e$ constraint release influenced the chain motion. Reptation scaling of the diffusion coefficient ($D \sim M^{-2}$) was also verified in [28] in the regime $M/M_e > 10$. Measurements of the self-diffusion coefficients in semi-dilute polymer solutions, by dynamic light scattering, were carried out in [1], and scaling laws predicted by the reptation were confirmed. Other studies that have verified the predictions from the reptation model can be found in [2, 10] (Fig. 1.11).

In Fig. 1.12, the motion of a chain through knots formed by surrounding chains is shown. For the time scale of motion of this chain, we can assume that "lock 1" and "lock 2" are the entanglements that do not vanish. Furthermore, based on entropic grounds, one can show that "hair pin" processes (referred to the conformation of a chain similar to the shape of a hair pin), as shown in Fig. 1.13, are unlikely and,

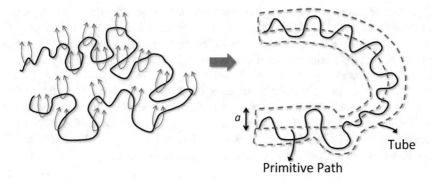

Fig. 1.11 Motion of a polymer chain through "knots" in a tube and its primitive path

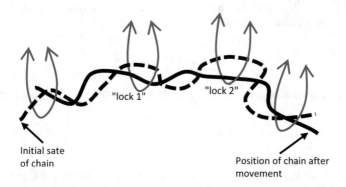

Fig. 1.12 Motion of a polymer chain through "knots"

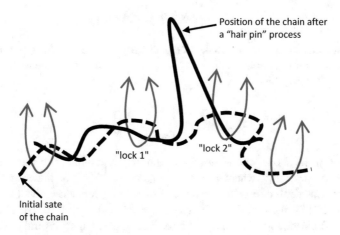

Fig. 1.13 Unfavorable "hair-pin" processes

therefore, the motion of a chain is constrained due to surrounding entanglements. For the chain shown in Fig. 1.13, to disengage from lock 2, a very large "hair pin" will be required. This concept can be understood by saying that the motion of a chain starts at the chain ends. In a later section, we shall estimate the probability of a "hair pin" process.

Now, we can estimate quantities such as the diffusion coefficient and timescales for motion during reptation. If we treat each monomer as an ideal coil with the diffusion coefficient of a single monomer as $D_{monomer}$, and a total of N monomers per chain, then the diffusion coefficient of the chain, D_{chain} (which contains N monomers, each with diffusion coefficient $D_{monomer}$) is given by

$$D_{chain} = \frac{D_{monomer}}{N}. \tag{1.18}$$

Similarly, if μ denotes the mobility inside a tube for one monomer, then the mobility of the chain is given as

$$\mu_{chain} = \frac{\mu}{N}. \tag{1.19}$$

Now, the time required for complete renewal of the tube (τ_{renew}) is estimated as

$$\tau_{renew} = \frac{L^2}{D_{chain}}, \tag{1.20}$$

here $L \sim N$, where we know $R_g \approx N^{1/2}$, and the key point is that the chain must move along its entire contour, and that is also the time scale for the "tube renewal". Hence,

$$\tau_{renew} \approx N^2 \times N = N^3. \tag{1.21}$$

Next, we can consider the translational diffusion. If in a time $\tau_t = \tau_{renew}$ a reptating chain moves along its tube by a length $L \approx Na$, and the center of mass is displaced by $N^{1/2}a$, then

$$D_{reptation} \approx \frac{R_g^2}{\tau_{renew}} \sim \frac{1}{N^2}. \tag{1.22}$$

Equation 1.22 shows the scaling law through reptation. The reptation time $\tau \sim N^3$ or, $\tau = \tau'N^3$, where τ' is of the order 10^{-10} s. For linear viscoelastic melts if we consider the relaxation time according to the reptation model $\tau = \tau'N^3$, with $\tau' = 10^{-10}$ s, then for $N = 10^5$ we get $\tau = 10^{-10} \times (10^5)^3 = 10^5$ s, which is quite long. Thus, the reptation time can increase and reach up to several minutes [7]. This also explains the long times needed for polymer adhesion. Also, it is exactly the mechanistic difference in reptation-like motion that could lead to interdiffusion and adhesion in comparison with plasticity-induced molecular mobilization and bonding, where bonding can occur in the order of 1 s (as will become clear later).

The displacement behaviors of chain segments during a diffusive process are noteworthy. At short diffusion times, the average displacement of the chain segments is less than or equal to the radius of gyration of the molecule, and deviations from classical Fickian diffusion are expected. For very short times, up to entanglement time (τ_e), when no entanglement effects are encountered, the segments do not feel constraints of the surrounding chains and $\langle r^2 \rangle^{1/2} \sim t^{1/4}$. At τ_e the displacement is given approximately by the diameter of the tube. For $\tau_e \leq t \leq \tau_R$ the movement perpendicular to the tube is restricted by entanglements and $\langle r^2 \rangle^{1/2} \sim t^{1/8}$. At the Rouse relaxation time τ_R, the movement of segments of the whole chain becomes correlated and $\langle r^2 \rangle^{1/2} \sim t^{1/4}$ for $\tau_R \leq t \leq \tau_{reptation}$. For times longer than the reptation time, normal Fickian diffusion is expected with $\langle r^2 \rangle^{1/2} \sim t^{1/2}$. For details and derivations the reader is referred to [4, 8, 22, 26].

We can also deduce the reptation time ($\tau_{reptation}$) or tube renewal time (τ_{renew}) from stress-relaxation experiments as follows. If there is an instantaneous entropic elastic stretch on the network, and subsequently the stretched chains are allowed to relax due to monomer, segmental, or other motions then, when the chains have moved a distance of $\approx R_g$ in time $\tau_{reptation}$, a large drop in the "stress" or "storage modulus" is noted.[1] Also, if E is the instantaneous elastic modulus of a system then $\eta = E\tau_{renew}$, or $\eta \approx \tau_{renew} \approx M^3$. Thus, reptation predicts the dependence of viscosity on molecular weight to the power of three. Doi and Edwards applied the reptation model to the rheology of entangled polymer liquids in 1978, and predictions of Doi-Edwards theory are in reasonable agreement with dynamic-mechanical measurements in entangled polymer melts. The dependence of viscosity on molecular weight based on Rouse and reptation models is shown schematically in Fig. 1.14.

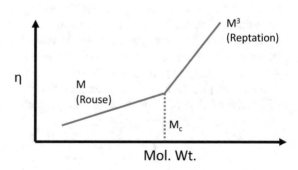

Fig. 1.14 Dependence of viscosity based on the Rouse model and reptation model

[1] The characteristic times for such large scale conformational rearrangements lie in the "terminal" region of the relaxation spectrum, say during stress relaxation experiments, and are very sensitive to chain length.

To conclude the discussions presented in the previous two sections, we empha-size that the stress-relaxation experiments and their characteristic behaviors, which are used to determine or estimate the Rouse (or reptation) times, and are conducted on polymer melts (or solutions). The notion of Rouse, Zimm or reptation relaxation dynamics is not directly applicable to solids. To measure them, there must be a polymer melt or solution. For the sake of completeness and clarity, we emphasize that the glassy modulus describes the linear elastic (instantaneous) response of the polymer at a temperature well below its T_g. At very high frequencies or timescales shorter than relaxation times, in a given polymer melt, an instantaneous elastic response will be observed. However, the polymer melt will eventually relax through a viscous flow. The reason why a polymer melt (liquid) can behave like a glass at high frequencies (or short time scales) is because the chain segments (or monomers) do not have sufficient time to relax the stress. Finally, the self-diffusion coefficient in polymer melts is strongly dependent on the temperature, molecular weight, and molecular characteristics. In the next section we quantitatively review some measured diffusion coefficients.

1.4 Diffusion in Polymer Melts

The goal of this section is to make a quantitative note of self-diffusion coefficients, primarily in polymer melts. Among other factors, adhesion due to interdiffusion in polymer melts is strongly a time dependent process, which directly depends on the self-diffusion coefficient of the polymer at a given temperature. Usually the diffusivities of polymer melts are quite small compared with the liquids composed of simple molecules (e.g., D_{water} at 298.1 K and 1 atm is approximately 2.3×10^{-9} m^2/s [19]; estimates of the diffusivities of other small molecular liquids are given in [20]). Therefore, adhesion due to interdiffusion in polymer melts has been reported to take time on the order of several minutes to hours. Typical values of self-diffusion coefficients from the literature are listed in Table 1.1. For a critical survey of the literature data on tracer, self-diffusion, and rheological measurements for chain dynamics in entangled polymers see [28].

As an example, if we set $x = 10$ nm then, for linear H-PB at 398.15 K, choosing a diffusivity of $D = 10^{-16}$ m^2/s, the time t required to diffuse this distance is

$$t = \frac{x^2}{2D} = \frac{100 \times 10^{-18}}{10^{-16}} = 1 \text{ s.}$$

Based on the above calculations, interdiffusion (up to 10 nm) across a polymer interface can occur in time on the order of a second. However, this is true for a polymer above its melting point. In the next chapter, we shall see that self-diffusivities in a glassy polymer can become extremely small; with D approaching the order of 10^{-24} m^2/s near T_g itself. Therefore, interdiffusion can take very long times as the temperature is brought closer to T_g from above. Well below T_g, no

Table 1.1 A short summary of the self-diffusion coefficient (D) from the literature. T_m and T_g stand for the melting and glass transition temperatures, respectively

Ref.	Polymer	Mol. Wt. (g/mol)	Temp. (K)	D (m^2/s)	Comments
[3]	Linear H-PB	5×10^4–20×10^4	398.15	10^{-14}–5×10^{-16}	$T_m \sim 381.15$ K
[9, 12]	Polyisoprene	560–9.82×10^4	373.15	2.2×10^{-10}–1.0×10^{-14}	$T_g \sim 192$ K [25]
[12]	Polybutadiene	$690 \times 4.99 \times 10^4$	373.15	7.0×10^{-11}–2.0×10^{-14}	$T_g \leq 183.15$ K
[23]	Polyethylene	200–12×10^4	448.15	6.6×10^{-10}–1.3×10^{-14}	$T_m \sim 353.15$–400.15 K
[17]	PDMS	500–5×10^5	293.65	7.0×10^{-10}–5×10^{-15}	$T_g \sim 150.15$ K
[11]	PS	600–19×10^3	487.8	2.5×10^{-10}–1.5×10^{-13}	$T_g \sim 333.15$–373.15 K

appreciable interdiffusion can be achieved on accessible experimental timescales. In another example, the interface between two half spaces of water molecules at ambient temperature, if brought into intimate contact, will disappear, i.e., diffusion on the order of 1 Å would be noted in time of 10^{-10} s (diffusivity value of $D_{water} = 2.32 \times 10^{-9}$ m^2/s is chosen).

In the following chapters, we shall clearly see that the mobility of polymers well below their T_g is extremely restricted, so that negligible diffusive motion can occur on experimental timescales. And, it is to this end when solid-state glassy (or semi-crystalline) polymers are plastically deformed there is an enhanced molecular-level mobility, sometimes comparable with that occurring in polymer melts.

References

1. E.J. Amis, C.C. Han, Cooperative and self-diffusion of polymers in semidilute solutions by dynamic light scattering. Polymer **23**(10), 1403–1406 (1982)
2. M. Appel, G. Fleischer, Investigation of the chain length dependence of self-diffusion of poly (dimethylsiloxane) and poly (ethylene oxide) in the melt with pulsed field gradient nmr. Macromolecules **26**(20), 5520–5525 (1993)
3. C.R. Bartels, B. Crist, W.W. Graessley, Self-diffusion coefficient in melts of linear polymers: chain length and temperature dependence for hydrogenated polybutadiene. Macromolecules **17**(12), 2702–2708 (1984)
4. F. Brochard-Wyart, P.-G. de Gennes, Hindered interdiffusion in asymmetric polymer-polymer junctions. Macromol. Symp. **40**(1), 166–177 (1990)
5. P.-G. de Gennes, Reptation of a polymer chain in the presence of fixed obstacles. J. Chem. Phys. **55**(2), 572–579 (1971)
6. P.-G. de Gennes, *Scaling Concepts in Polymer Physics* (Cornell University Press, London, 1979)
7. P.-G. de Gennes, L. Leger, Dynamics of entangled polymer chains. Annu. Rev. Phys. Chem. **33**(1), 49–61 (1982)
8. M. Doi, S.F. Edwards, The Theory of Polymer Dynamics (Clarendon, Oxford, 1986)
9. M. Doxastakis, D. Theodorou, G. Fytas, F. Kremer, R. Faller, F. Müller-Plathe, N. Hadjichristidis, Chain and local dynamics of polyisoprene as probed by experiments and computer simulations. J. Chem. Phys. **119**(13), 6883–6894 (2003)
10. G. Fleischer, Self diffusion in melts of polystyrene and polyethylene measured by pulsed field gradient nmr. Polym. Bull. **9**(1–3),152–158 (1983)
11. G. Fleischer, Temperature dependence of self diffusion of polystyrene and polyethylene in the melt an interpretation in terms of the free volume theory. Polym. Bull. **11**(1), 75–80 (1984)
12. G. Fleischer, M. Appel, Chain length and temperature dependence of the self-diffusion of polyisoprene and polybutadiene in the melt. Macromolecules **28**(21), 7281–7283 (1995)
13. P.J. Flory, *Principles of Polymer Chemistry* (Cornell University Press, Ithaca, 1953)
14. W.W. Graessley, Entangled linear, branched and network polymer systems–molecular theories, in *Synthesis and Degradation Rheology and Extrusion* (Springer, New York, 1982), pp. 67–117
15. H. Hervet, L. Leger, F. Rondelez, Self-diffusion in polymer solutions: a test for scaling and reptation. Phys. Rev. Lett. **42**(25), 1681 (1979)
16. W. Hess, Self-diffusion and reptation in semidilute polymer solutions. Macromolecules **19**(5), 1395–1404 (1986)
17. R. Kimmich, W. Unrath, G. Schnur, E. Rommel, NMR measurement of small self-diffusion coefficients in the fringe field of superconducting magnets. J. Magnet. Reson. (1969) **91**(1), 136–140 (1991)

18. J. Klein, Evidence for reptation in an entangled polymer melt. Nature **271**(5641), 143–145 (1978)
19. K. Krynicki, C.D. Green, D.W. Sawyer, Pressure and temperature dependence of self-diffusion in water. Farad. Discuss. Chem. Soc. **66**, 199–208 (1978)
20. J. Matthiesen, R.S. Smith, B.D. Kay, Mixing it up: measuring diffusion in supercooled liquid solutions of methanol and ethanol at temperatures near the glass transition. J. Phys. Chem. Lett. **2**(6), 557–561 (2011)
21. T. McLeish, Tube theory of entangled polymer dynamics. Adv. Phys. **51**(6), 1379–1527 (2002)
22. J.T. Padding, Theory of polymer dynamics. http://wwwthphys.physics.ox.ac.uk/people/ArdLouis/padding/PolymerDynamics_Padding.pdf
23. D. Pearson, G. Ver Strate, E. Von Meerwall, F. Schilling, Viscosity and self-diffusion coefficient of linear polyethylene. Macromolecules **20**(5), 1133–1141 (1987)
24. P. Rouse, A theory of the linear viscoelastic properties of dilute solitions of coiling polymers. J. Chem. Phys. **21**, 1272 (1953)
25. Segmental and chain dynamics in amorphous polymers. http://comse.chemeng.ntua.gr/segmdyn_polpage.htm
26. I. Teraoka, *Frontmatter and Index* (Wiley Online Library, Hoboken, 2002)
27. J. Von Seggern, S. Klotz, H. Cantow, Reptation and constraint release in linear polymer melts: an experimental study. Macromolecules **24**(11), 3300–3303 (1991)
28. S.-Q. Wang, Chain dynamics in entangled polymers: diffusion versus rheology and their comparison. J. Polym. Sci. Part B Polym. Phys. **41**(14), 1589–1604 (2003)
29. R.J. Young, P.A. Lovell, *Introduction to Polymers* (CRC Press, Boca Raton, 2011)
30. B.H. Zimm, Dynamics of polymer molecules in dilute solution: viscoelasticity, flow birefringence and dielectric loss. J. Chem. Phys. **24**(2), 269–278 (1956)
31. D. Masao, S.F. Edwards, Dynamics of concentrated polymer systems. Part 1.—Brownian motion in the equilibrium state. J. Chem. Soc., Faraday Trans. 2 **74**, 1789–1801 (1978)

Chapter 2
Structure and Properties of Polymer Glasses

Abstract In the solid state, the long-chain molecules of a polymer can exist in a randomly oriented (disordered) amorphous state or a partially ordered semi-crystalline state. Regions of amorphous and crystalline domains are interdispersed throughout the bulk. In this chapter, we discuss the aspects of kinetics of motions in amorphous polymers, and show that the mobility of polymer molecules in the solid state is severely restricted. Understanding of the concepts of glass transition, glass formation, and the nature of a glassy structure is developed here. The main goal is to clearly highlight that polymer molecules below the T_g are kinetically trapped. Hence, establishment of polymer adhesion due to interdiffusion well below the T_g, on short experimental timescales is not feasible. We also discuss key concepts of lowering the T_g on free surfaces of polymers, as is reported commonly.

2.1 Polymer Glasses

Subtleties aside, if polymeric liquids are cooled they can partially crystallize (i.e., form a locally ordered solid state) or form a glass (disordered solid). Crystalline domains are associated with a sharp melting temperature (T_m), whereas the temperature at which the transition from liquid to a glass occurs is known as the T_g. Melting (or crystallization) is a phase transformation and a first-order transition (marked by a sharp change in a physical property). However, during glass formation, there is usually some non-uniform change in the property; especially marked by an extreme slowdown of kinetics [23]. A glass is not a (supercooled) liquid because a supercooled liquid is formed below the melting point of a crystallizable liquid if there are no nucleation sites to initiate crystallization (Fig. 2.1).

Understanding the origin of the extraordinary slowdown of relaxation processes during glass formation has been one of the main challenges in the physics of glasses for scientists and researchers. Philip W Anderson (a Nobel Prize-winning physicist at Princeton University in 1995) said: "The deepest and most interesting unsolved problem in solid state theory is probably the theory of the nature of glass and the glass transition. This could be the next breakthrough in the coming decade. The

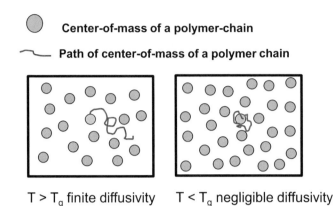

Fig. 2.1 "Diffusion" above T_g and below T_g for glassy materials

solution of the problem of spin glass in the late 1970s had broad implications in unexpected fields like neural networks, computer algorithms, evolution, and computational complexity. The solution of the more important and puzzling glass problem may also have a substantial intellectual spin-off. Whether it will help make better glass is questionable." It has been stated that most of the water that exists in the universe is in a glassy state [37]. More recent commentary on the elusive nature of glassy dynamics can be found in [15, 34]. A good source of reference on disordered glassy media is [8].

From a thermodynamic viewpoint, the glassy state is unstable. That is, for a single-component glass at constant temperature and pressure, the Gibbs free energy will not be at a minimum. In contrast to a supercooled liquid, a glass is continuously relaxing, possibly too slowly to measure, towards a more stable state (i.e., a local free-energy minimum). If experimental observations are made on timescales shorter than those of the molecular motions which allow the glass to relax, then the glass is mechanically stable for practical purposes, even though it is thermodynamically unstable. We can also think about the complex behavior of such liquids, or even a glass in terms of a potential energy hyper-surface in higher dimensions, with its elemental configurations as coordinates. In a glass, the system clearly resides within a local minimum (or basin) on this hyper-surface, and executes collective vibrational modes of motion in accordance with its basin shape. Below the T_g, for example, during annealing, the system can slowly explore nearby lower energy minima while lowering its entropy. At very high temperatures, it sees no "energy wells" and can readily sample new configurations. We review the concepts of free-volume theory and diffusion [27], which are useful for gaining a basic understanding of glass transition.

2.2 Free Volume Theory and Diffusion

As the temperature of a glass-forming liquid is lowered, its specific volume and free volume decreases. The different regimes during cooling (crystallization temperature, supercooled liquid, and glass transition) are shown in Fig. 2.2. Free volume is the "elbow room" or space available for molecules to sample new configurations and move spatially. As the temperature is lowered, molecular motion is slowed down and, due to the insufficient time to rearrange into an equilibrium configuration, the system cannot attain its equilibrium specific volume at that temperature. The experimentally observed specific volume starts deviating from the equilibrium value. The glass transition does not occur immediately but over some range of temperature, and this is called the "transformation range". Thus, glass transition is not a phase transformation but a kinetic event, and the cooling rate has an effect on the thermodynamic (and dynamic) properties of the glass.

Cohen and Turnbull [17] were the first to introduce the notion that molecular transport is controlled by percolation of the free volume. They derived the dependence of the diffusion coefficient for a liquid assumed to comprise hard spheres. We present some basic ideas behind the free volume theory. Let the volume of each hard sphere be denoted by V^*. Then, the volume occupied in a liquid containing N spheres $= NV^*$. If V_f is free volume per sphere, then the total volume of the liquid $V_{total} = N(V_f + V^*)$. The migration rate, or self-diffusion, of a hard-sphere is proportional to the probability of finding a hole of volume V^* (or larger) adjacent to the sphere. If free-volume voids become available to spherical molecules by means of natural thermal fluctuations within a liquid, the molecules undergo a single step in the diffusion process. By describing the dispersion of free-volume elements within a liquid in mathematical terms, Cohen and Turnbull developed a distribution that provided the probability of finding a free volume hole of a specific size. Accordingly, the diffusion coefficient, considered proportional to the probability of

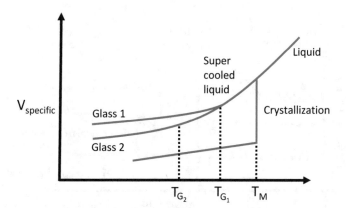

Fig. 2.2 Specific volume and glass transition state. An example of a liquid, cooled at different rates, resulting in different states (glassy or crystalline)

finding a hole of volume V^* or larger, is given as

$$D = Ae^{(-\gamma V^*/V_f)}.$$

Here, V^* is the minimum volume hole size into which a molecule can jump, and γ is a numerical factor which lies between 0.5 and 1.

Let $V(0)$ be the specific volume of the liquid at 0 K, and V be the volume at some temperature T. The specific volume $V(0)$ is assumed to be independent of molecular weight for polymeric liquids. Accordingly, the free volume of the liquid (at temperature T) is given as

$$V_f = V - V(0).$$

As the temperature is increased from 0 K, the increase in free volume is realized partly by the homogeneous expansion of the material due to the increasing amplitude of harmonic vibrations with temperature, and partly by the formation of holes or vacancies which are distributed discontinuously throughout the material at any instant. Thermal expansion in the glassy and crystalline states is similar. In each case, expansion is dominated by atomic vibrations which are very similar in the two states. The first type of free volume is the interstitial free volume (V_{fL}); this is associated with molecules and increases homogeneously in the liquid. The second type is called hole free volume V_{fH}, and it is implicitly assumed that it can be distributed without any energy expense. Thus,

$$V_f = V_{fL} + V_{fH}.$$

The entrapped free volume during cooling and glass formation are shown schematically in Fig. 2.3. Free volume decreases during cooling from the rubbery state until the glass transition temperature is reached, at which point it attains a crit-

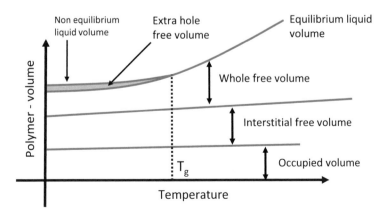

Fig. 2.3 Kinetic arrest and the glass transition temperature

ical minimum, and molecular rearrangement is effectively frozen. In a temperature range well below or near the T_g, dynamics are expected to become increasingly cooperative. As the temperature is lowered, specific volume continues to decrease as temperature decreases; however, the thermal expansion coefficient in the glassy state is significantly smaller than in the liquid or in the supercooled liquid states. At temperatures below T_g the polymer tries to reduce its free volume and enthalpy but the dynamics become extremely sluggish.

The free volume model used to describe the behavior of glass-forming liquids has limitations because of several reasons quoted in [22]. Mode-coupling theory and (entropic) Adam and Gibbs theory are among other earlier theories which have been employed to explain the behavior of glass and glass transition [26]. Modern theories include random first order transition (mosaic theory), spin glasses theory: mean-field p-spin model, frustration-limited domains, hierarchical random energy model, dynamical facilitation theory, free-energy landscape theories, and two-temperature thermodynamic theory. For an overview on these theories, detailed references can be found in [7]. Cooperative motion near the T_g is a key part of recent theories of the glass transition. Detailed discussions on these topics are beyond the scope of this monograph, and interested readers are referred to [43].

At high temperatures, $T > T_g$, the system is *ergodic* and molecules can sample new orientations [3]. In the regime $T \sim T_g$, the system is non-ergodic except for very long annealing times, and hence has time-dependent properties (annealing and ageing). During ageing, the change in molecular properties occurs over time because of the very slow molecular arrangements in glass.

For ordinary liquids, the Stokes-Einstein relation between the diffusion coefficient and viscosity is valid at high temperatures. Deviations occur for $T < T_c$, where T_c is the critical temperature in the supercooled regime [56]. Below T_c, activated "hopping" processes are believed to govern the dynamics. This also implies that diffusion in liquids occurs by different mechanisms at high and low temperatures. At low temperatures, molecules move by crossing substantial potential barriers (i.e., activated transport or hopping), whereas, at high temperatures, thermal energies are comparable with barrier heights and translational motion have the characteristics of free diffusion. If free diffusion ceases (around the critical temperature T_c), then rearrangements become cooperative.

At temperatures well the below T_g, primary relaxation processes are effectively frozen but secondary relaxation processes can be active [39]. In the next section, we discuss these relaxation processes in detail.

2.3 Relaxation Processes and Glass Transition

In any glass-forming system, several relaxation mechanisms are possible. Bond-length and bond-angle vibrations can occur on timescales on the order of 10^{-13} to 10^{-14} s, which are relatively insensitive to molecular packing. However, correlated segmental relaxations have larger timescales. Two common relaxation processes

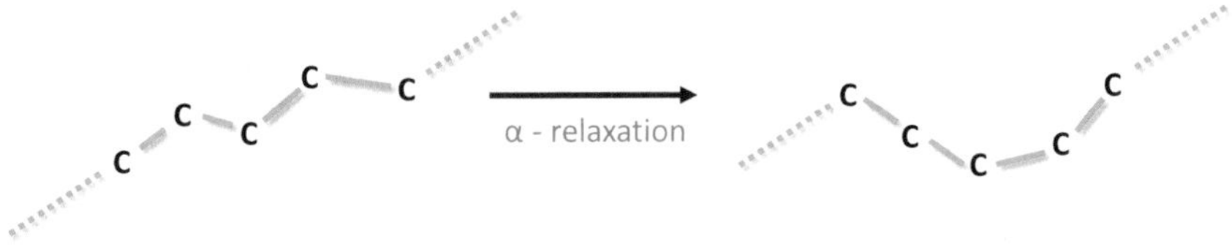

Fig. 2.4 Cooperative motion of chain segments as α-relaxation

worth discussing are the so-called α and β relaxations. An example of an α-relaxation is shown in Fig. 2.4, where typically 5–10 backbone bonds participate in one relaxation event. Vibrations of skeletal bonds in their energy wells and conformational transitions across torsional-energy barriers are accomplished through localized distortions of the chain back bones. The α-relaxation is active above or near the T_g. β-relaxations are usually rotations of pendant groups (simple flips of rings or side groups), and are active down to temperatures below the T_g. Other lower-order relaxation mechanisms can be active below the T_g. However, only at temperatures well above the T_g full Rouse-type relaxations are possible. Well below the T_g, large-scale conformational rearrangement, reorientation, and self-diffusion of chains are frozen.

One of the best-known signatures of a glassy behavior is the stretched-exponential relaxation (SER) law (see [28] for discussions), according to which the time-dependent decay of many types of perturbations looks like

$$f_\beta(t) = e^{-(t/\tau)^\beta},$$

instead of a simple exponential (with $\beta = 1$). Here, the timescale τ is the characteristic relaxation time constant, and the exponent β may be substantially less than unity. As the temperature drops, segmental relaxations slow down dramatically and motions become increasingly cooperative, and viscous liquids close to the T_g exhibit non-exponential relaxation. The temporal behavior of the response function, for example, viscoelastic relaxation [42], dielectric relaxation [52, 66], glassy relaxations [14], relaxation in polymers [40, 45], or polymer melts [63, 64], can often be described by the stretched exponential function, also popularly known as the Kohlrausch-Williams-Watts (KWW) function. There are two ways to interpret the stretched exponential relaxation:

(i) One can imagine a heterogeneous set of environments and that relaxation in a given environment is nearly exponential, but the relaxation times vary significantly among environments.

(ii) One can consider that the whole system is homogeneous but relaxes non-exponentially.

Importantly, the dynamic heterogeneity is one of the central themes of glass-transition, and the length scale of this heterogeneity is called the "characteristic length". For a wide range of molecular glass-formers near the T_g, this has been shown to be typically of the order 1.0–3.5 nm in length [35]. The temperature dependence of the characteristic relaxation time τ can be Arrhenius or non-Arrhenius. As cooling is carried out and the T_g is approached, two types of behaviors are noted and liquids are categorized accordingly: (i) "Strong liquids", which show Arrhenius-like behavior for viscosity and relaxation time, and have a three dimensional network structures comprising covalent bonds. (ii) "Fragile liquids", which demonstrate non-Arrhenius relaxation properties and typically consist of molecules interacting through non-directional and non covalent interactions. The non-Arrhenius temperature dependence of relaxation or viscosity can be described by the Vogel-Tammann-Fulcher (VFT) equation as

$$\tau = \tau_0 exp(\frac{\beta}{T - T_\infty}).$$

Here, T_∞ is a reference temperature, and if $T_\infty = 0$, the behavior is Arrhenius. Williams-Landel-Ferry (WLF) [67] is another model describing this behavior and is mathematically equivalent to VFT. In the context of polymers, the primary glassy to rubbery transition is often accepted to be better represented by the WLF equation.

2.4   Examples of Slow Kinetics in Glasses

If the temperature of a glass-forming liquid is lowered and the T_g is approached from above, the kinetics of a glass-forming system show a drastic slowdown, timescales for relaxations become significantly large [1, 18, 23, 28, 36, 38, 60], and the system is completely frozen out with respect to cooperative segmental (or α-like relaxations) [3] on accessible experimental times. For example, the glass transition state itself is typically characterized by viscosity and diffusivity values of 10^{13} Poise and 10^{-24} m^2/s, respectively [59].

In molecular liquids near the T_g, it may take minutes-to-hours for a molecule of diameter less than 10 Å to reorient [28]. In [62], the self-diffusivity of a single-component glass former, *tris*-naphthylbenzene (TNB), fell from $D = 10^{-14}$ m^2/s at 405 K to $D = 10^{-20}$ m^2/s as the T_g (close to 335 K) was approached. A slowdown of self-diffusion, with D approaching 10^{-20} m^2/s near the T_g, was noted for O-terphenyl and a metallic melt of Pd–Cu–Ni–P systems [55]. In [48], it was shown that as the T_g was approached from above, the diffusivity for O-terphenyl (OTP) approached 10^{-20} m^2/s. For O-teraphenyl, the rotation time at the T_g was about 10^4 s. This is an astoundingly long time compared with the pico-second or nano-second rotation times observed in typical liquids. A pure sample liquid such as O-teraphenyl will not crystallize for years in a test tube at room temperature at, for example, 35 K below T_m. For actactic polymers (i.e., polymers with random stereo chemistry), often the crystalline state is never obtained and may be higher in free energy than the liquid state at all temperatures [29]. For example, in [58], the authors estimated the temperature below the T_g for amorphous systems; the timescales for molecular motion (i.e., relaxation time) exceeded the expected lifetimes or storage times of pharmaceutically important glasses. As noted in [36], volumetric equilibrium in polystyrene (PS) is achieved within a "reasonable" timescale of 100 h for ageing temperatures only down to about 10 °C below the T_g, (taking T_g = 105 °C). Also see [18] for detailed discussions. In [16], it was shown that the relaxation time at the T_g for polymethyl methacrylate (PMMA) was 100 s.

In [68], while discussing "Do cathedral glasses flow?", Zanotto points out that typical window glasses which contain K_2O-Na_2O-CaO-MgO-$Al_2O_3 \cdot SiO_2$ and a certain amount of impurities may flow appreciably at ambient temperatures only over geological periods, not within limits of human history. A numerical calculation of the relaxation timescales at ambient temperature was made, and a characteristic flow time was found to be 10^{32} years, whereas the life of our universe is estimated to be $\approx 10^{10}$ years. Also see [25] for more discussions. Another piece of evidence of extremely large time scales for flow in the glassy state, at room temperature, stems from the fact that glassware from thousands of years ago remain undeformed in museums worldwide.

Based on the above discussions it is clear that glassy polymers, well below their T_g, will not exhibit any long-range diffusive motion on their own on experimental timescales. Therefore, adhesion through interdiffusion on short timescales is not possible.[1] However, researchers have reported a lowering in the surface T_g compared with the bulk T_g, and the plausible role of lowering of surface T_g in bonding of glassy polymers, through interdiffusion, below their bulk T_g. The phenomenon of lowering of T_g has also been reported in "thin" polymer films (where thickness is of the order of tens of nano-meters). In the next section, we discuss surface-related lowering of the T_g in polymers, and clarify that, in spite of enhanced mobility on the free surfaces of the polymers at temperatures well below the bulk T_g, adhesion in timescales on the order of 1 s is not possible unless accompanied by bulk plastic deformation.

2.5   Low T_g on the Free Surface of Polymers

A rubber-like layer on the free surface of glassy polymers has been known for sometime. This phenomenon can have several technological consequences when polymer free surfaces or interfaces are involved in contact interactions. In [50], the authors studied the morphology produced due to the interaction of the atomic force microscope (AFM) tip with the PS surface. They reported, for the first time, that the morphology produced on the nanometer scale was dependent upon the molecular weight of the polymer, and inferred that the surface of PS behaved more like a material exhibiting rubber elasticity than one in a glassy state. More

---

[1] Well below T_g implies several tens of °C below the bulk T_g.

recently, Fakhraai et al. [30] studied the time-dependent relaxation of the PS surface by partially embedding and removing gold nanospheres. Surface relaxation was observed at all temperatures and provided strong evidence of enhanced mobility on the surface. However, they reported that relaxation times associated with such enhanced mobility were closer to those of β-relaxation processes.

An explanation for T$_g$ lowering, due to chain end localization at the free surface, was provided by Mayes [49]. Chain ends are associated with greater free volume compared with that of other segments of the chain; thus segregation of chain ends at a free surface leads to effective lowering of the molecular weight on the free surface of the polymer. This, along with the Fox-Flory relationship; ($T_g = T_{g,\infty} - K/M_n$, which gives the dependence of the T$_g$ on the molecular weight M$_n$ of the polymer), explains lowering of the T$_g$ on the free surface of the polymer due to chain end segregation.[2] It should be noted that Mayes assumed the localization of chain ends and explained the lowering of T$_g$ on the free surface; however, the localization of the chain ends is not always necessary, and depends on several factors as discussed next.

A simulation study on dense polymer melt between two hard walls [5] reported that the effects of loss in configurational entropy due to the presence of a hard wall (i.e., no interaction) lead to a force that drives the polymers away from the interface into the bulk like matrix, but this is encountered with competition due to the packing constraint. The simulations revealed that the density profiles of monomers and of end monomers were enhanced at the walls.

De Gennes [21], based purely on physical grounds, argued that an ideal monomer experiences an entropy loss of the order K_BT. The entropy of an ideal Gaussian monomer coil scales with K_BT, and an ideal monomer (treated as a coil) will suffer entropy loss near the free surface, so the chain-end segregation is favored on entropic grounds. Illustration of this idea is shown in Fig. 2.5, where "not possible" configurations are due to rigid reflection, and entropically "favorable" and "less favorable" configurations are shown. If monomers are away from the free surface, then they can sample more orientations, such distributions are favored on entropic grounds. Figure 2.6 provides an example of a monomer constrained to lie flat near the surface and, based purely on entropic grounds, this is unfavorable.

De Gennes provided three regimes for chain-end segregation by considering the difference between the surface tension of an infinite polymer chain (γ_∞) and that of end groups (γ_{end}), and comparing a quantity $\delta a^2 = (\gamma_\infty - \gamma_{end})a^2$ (where a represents the segment length) with K_BT. If δa^2 is much smaller than the thermal energy (K_BT), the first regime, then the surface-end concentration should be equal to that of the bulk. If the ratio of δa^2 and K_BT is near unity, the second regime, then all chain ends within a distance of the end-to-end vector should localize on the surface. Finally, when δa^2 is much larger than K_BT, the last regime, then chains would stretch to accommodate more chain ends. A simple understanding

---

[2]The relaxation modes marking the onset of a rubbery behavior on the free surface could be different than the relaxation modes characterizing the bulk T$_g$.

Fig. 2.5 Free surface of the
polymer and segregation of
the chain end

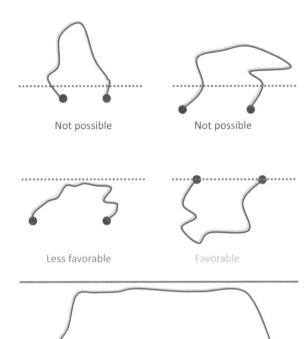

Fig. 2.6 Free surface of the
polymer and monomer lying
close to the surface

on this behavior could be gained as follows. The first regime $\frac{\delta a^2}{K_B T} \ll 1$ could
imply that ends have high energy cost (say δ is small because γ_{end} is high) to be
placed on the surface, but this is also favorable to be placed on entropic grounds, so
preferential segregation does not occur. In the second regime, we could interpret
that the energy expense to place the ends on the surface is low and, hence, on
entropic grounds, chain ends localize on the surface. In the last regime, there is
almost no energy expense to place the ends on the surface, and, hence a higher
concentration of ends would be strongly favorable. Thus, the resultant effect of
entropic and enthalpic factors can lead to segregation or repulsion of chain ends at
the free surface. However, such effects decay within distances comparable with the
radius of gyration of the polymer. The polymeric films discussed in this monograph
comprise a base polymer and a compatible plasticizer (small molecule with only a
few segments). Again, based purely on entropic grounds, we could speculate that
the smaller molecule is likely to suffer less entropy loss compared with polymer
chains on the free surface and, therefore, may segregate (Fig. 2.7). Absorption and
adsorption of foreign molecules in the solvent-cast polymer films could also play a
part in affecting their surface properties and, thus, bonding.[3]

---

[3]Absorption is the process in which a fluid is dissolved by a liquid or a solid (absorbent).
Adsorption is the process in which atoms, ions or molecules from a substance (gas, liquid or
dissolved solid) adhere to a surface of the adsorbent. Adsorption is a surface-based process in
which a film of the adsorbate is created on the surface, while adsorption involves, the entire volume
of the absorbing substance.

Fig. 2.7 Excess localization of chain ends at the free surface of a glassy polymer

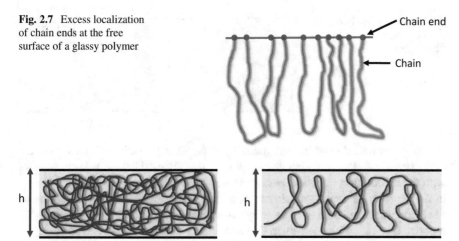

Chain end

Chain

Fig. 2.8 Confined "thin" polymer film and a single chain

The presence of a rubber-like layer, at the free surface of a glassy polymer, with enhanced dynamics has also been shown in computer simulations [47]. The mean configuration of the macromolecules at the free surface is also perturbed in the direction normal to the surface. However, such effects decay within distances comparable with the radius of gyration of the polymer. For a detailed review see [33].

We emphasize that lowering of the T_g (with respect to bulk T_g) on the free surface of a polymer with glassy bulk beneath is a different issue than lowering of the T_g (with respect to bulk T_g) in "thin" films. An example of a "thin" confined film is shown in Fig. 2.8. Fundamentally, one must consider four aspects in determining the T_g of any such system: (i) the physical property being measured, from which glass transition will be inferred; (ii) the method of measuring the physical property of interest; (iii) sample geometry, physical and chemical properties of the material; and (iv) most importantly, the plausible molecular-level or segmental relaxation processes associated with the observed transition. An interplay of all these factors can lead to a wide range of results for "thin" films. For example, Prevosto et al. [51] reported an upper and lower glass transition, within free-standing PS films of thickness less than 70 nm, originating from two distinct and simultaneous mechanisms. These were segmental (α-) relaxation and the sub-Rouse modes with a longer length scale compared with those involved in a typical (α-) relaxation, but shorter than that of the Rouse mode. Dalnoki-Veress et al. [20] suggested a mechanism of mobility in thin freely standing films that is inhibited in the bulk and distinct from the usual cooperative motion near the T_g, and explained lowering of the T_g in such films. In Fig. 2.9, due to confinement of the chain segments in this narrowly confined space, segments lying close to the free surface could "slide" and exhibit more mobility than if embedded within a bulk. In the same study, the role of net attraction between the "thin" film and the substrate was emphasized in

Fig. 2.9 A single polymer
chain confined in a narrow
"thin" space

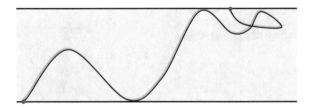

lowering or increasing of the measured T_g of the film. Lowering of the T_g and the effect of the substrate are also discussed in [31]. Other studies on lowering the T_g with decreasing film thickness are reported in [6, 41, 65]. In [54], a slight increase in the T_g with decreasing thickness of thin films of PMMA on the native oxide of silicon was noted, and the role of hydrogen bonding was highlighted. Reiter [53] reported lowering of the density in PS films thinner than the end-to-end distance and a resulting increase in the mobility. In [46], the authors conducted surface-relaxation studies on thin films (25–170 nm) made of PS and reported no evidence of enhanced mobility at the free surface. Similarly, in [32] T_g measurements for thin polymer films were conducted using atomic force microscopy (AFM), and reported that the T_g was independent of film thickness in the range 17–500 nm, strength of substrate interactions, or even the presence of a substrate.

Adhesion between two pieces of glassy polymers, due to interdiffusion of macromolecules, cannot be noted on experimental timescales at temperatures well below the T_g. This is because any long-range molecular mobility (or macro-molecular diffusivity) is essentially absent [2, 4, 24, 61] and cooperative segmental motions are inhibited (α-relaxation) [4], which can yield entanglements across the interface through interdiffusion. Even at the T_g of a polymer, the self-diffusivity of macromolecules is extremely small, and "sluggish" molecular kinetics prevail [19]. The T_g itself is typically characterized by viscosity and diffusivity values of 10^{13} Poise and 10^{-24} m^2/s, respectively [59]. Assuming a viscosity of 10^{13} Poise at the T_g [44], self-diffusion coefficients of forty polymers (at their respective glass-transition temperatures) were estimated to be approximately 10^{-25} m^2/s. Similarly, several other examples of extremely slow kinetics in glass-forming liquids near the T_g (marked by very small diffusion coefficients) have been reported in the literature [48, 55, 62]. Nevertheless, there have been reports of polymer adhesion below the T_g [9, 11–13] through application of mild pressure over experimental timescales of minutes and hours, and such studies have attributed bonding (below the T_g) to a rubber-like surface layer present on the free surface of the polymer, where the T_g is naturally lowered, and molecular mobility is enhanced [30, 33, 47, 50].

In the past two decades, researchers have reported achievement of polymer adhesion due to interdiffusion at temperatures below the bulk T_g using relatively long healing times of several minutes [12, 57], hours [10, 13], and even up to 1 day [9]. Such studies have claimed that bonding due to interdiffusion at temperatures below the bulk T_g is possible because the surface layer of a glassy polymer is in a rubbery state. In our assessment, such studies have completely ignored the role of mechanical activation imposed due to applied "pressure." Loading configurations claiming to apply pressure, as reported in the literature, lead to non-hydrostatic loadings (i.e., loadings with non-zero components of the stress deviator tensor), and this can cause creep or plastic deformation to cause enhanced mobility and interpenetration over a long contact duration.

References

1. A. Alegria, E. Guerrica-Echevarria, L. Goitiandia, I. Telleria, J. Colmenero. Alpha-relaxation in the glass transition range of amorphous polymers. 1. Temperature behavior across the glass transition. Macromolecules **28**(5), 1516–1527 (1995)
2. A. Alegria, E. Guerrica-Echevarria, L. Goitiandia, I. Telleria, J. Colmenero, Alpha-relaxation in the glass transition range of amorphous polymers. 1. Temperature behavior across the glass transition. Macromolecules **28**(5), 1516–1527 (1995)
3. C.A. Angell, K.L. Ngai, G.B. McKenna, P.F. McMillan, S.W. Martin, Relaxation in glassforming liquids and amorphous solids. J. Appl. Phys. **88**(6), 3113–3157 (2000)
4. C.A. Angell, K.L. Ngai, G.B. McKenna, P.F. McMillan, S.W. Martin, Relaxation in glassforming liquids and amorphous solids. J. Appl. Phys. **88**(6), 3113–3157 (2000)
5. J. Baschnagel, K. Binder, On the influence of hard walls on structural properties in polymer glass simulation. Macromolecules **28**(20), 6808–6818 (1995)
6. O. Bäumchen, J. McGraw, J. Forrest, K. Dalnoki-Veress, Reduced glass transition temperatures in thin polymer films: surface effect or artifact? Phys. Rev. Lett. **109**(5), 055701 (2012)
7. L. Berthier, G. Biroli, Theoretical perspective on the glass transition and amorphous materials. Rev. Mod. Phys. **83**(2), 587 (2011)
8. K. Binder, W. Kob, *Glassy Materials and Disordered Solids: An Introduction to Their Statistical Mechanics* (World Scientific, Singapore, 2011)
9. Y.M. Boiko, On the formation of topological entanglements during the contact of glassy polymers. Colloid Polymer Sci. **290**(12), 1201–1206 (2012)
10. Y.M. Boiko, Is adhesion between amorphous polymers sensitive to the bulk glass transition? Colloid Polymer Sci. **291**(9), 2259–2262 (2013)
11. Y.M. Boiko, Statistical adhesion strength of an amorphous polymer–its miscible blend interface self-healed at a temperature below the bulk glass transition temperature. J. Adhes. **96**(8), 760–775 (2020)
12. Y.M. Boiko, R.E. Prud'homme, Strength development at the interface of amorphous polymers and their miscible blends, below the glass transition temperature. Macromolecules **31**, 6620–6626 (1998)
13. Y.M. Boiko, V.A. Zakrevskii, V.A. Pakhotin, Chain scission upon fracture of autoadhesive joints formed from glassy poly (phenylene oxide). J. Adhes. **90**(7), 596–606 (2014)
14. R. Chamberlin, G. Mozurkewich, R. Orbach, Time decay of the remanent magnetization in spin-glasses. Phys. Rev. Lett. **52**(10), 867 (1984)
15. K. Chang, The nature of glass remains anything but clear. New York Times, 29 (2008)
16. K. Chen, K.S. Schweizer, Molecular theory of physical aging in polymer glasses. Phys. Rev. Lett. **98**(16), 167802 (2007)
17. M.H. Cohen, D. Turnbull, Molecular transport in liquids and glasses. J. Chem. Phys. **31**(5), 1164–1169 (1959)
18. R.H. Colby, Dynamic scaling approach to glass formation. Phys. Rev. E **61**(2), 1783 (2000)
19. N.R. Council, *Polymer Science and Engineering: The Shifting Research Frontiers* (The National Academies Press, Washington, DC, 1994)
20. K. Dalnoki-Veress, J. Forrest, P.-G. de Gennes, J. Dutcher, Glass transition reductions in thin freely-standing polymer films: a scaling analysis of chain confinement effects. Le J. de Phys. IV **10**(PR7), Pr7–221 (2000)
21. P.-G. de Gennes, in *Physics of Polymer Surfaces and Interfaces*, ed. by I.C. Sanchez, L.E. Fitzpatrick (Butterworth-Heinemann, Boston, 1992)

22. P.-G. de Gennes, On polymer glasses. J. Polym. Sci. Part B: Polym. Phys. **43**(23), 3365–3366 (2005)
23. P.G. Debenedetti, F.H. Stillinger, Supercooled liquids and the glass transition. Nature **410**(6825), 259–267 (2001)
24. P.G. Debenedetti, F.H. Stillinger, Supercooled liquids and the glass transition. Nature **410**(6825), 259–267 (2001)
25. Do cathedral glasses flow? http://www.doitpoms.ac.uk/tlplib/glass-transition/myth.php
26. A. Dobrynin, Glass and glass transition. http://faculty.ims.uconn.edu/~avd/class/2006/cheg351/lec6.pdf
27. J. Duda, J.M. Zielinski, Free-volume theory, in *Diffusion in Polymers*, ed. by P. Neogi (Marcel Dekker, Inc., New York, 1996)
28. M. Ediger, C. Angell, S.R. Nagel, Supercooled liquids and glasses. J. Phys. Chem. **100**(31), 13200–13212 (1996)
29. M.D. Ediger, Spatially heterogeneous dynamics in supercooled liquids. Annu. Rev. Phys. Chem. **51**(1), 99–128 (2000)
30. Z. Fakhraai, J.A. Forrest, Measuring the surface dynamics of glassy polymers. Science **319**(5863), 600–604 (2008)
31. J. Forrest, K. Dalnoki-Veress, J. Stevens, J. Dutcher, Effect of free surfaces on the glass transition temperature of thin polymer films. Phys. Rev. Lett. **77**(10), 2002 (1996)
32. S. Ge, Y. Pu, W. Zhang, M. Rafailovich, J. Sokolov, C. Buenviaje, R. Buckmaster, R. Overney, Shear modulation force microscopy study of near surface glass transition temperatures. Phys. Rev. Lett. **85**(11), 2340 (2000)
33. R.N. Haward, *The Physics of Glassy Polymers* (Springer Science & Business Media, Dordrecht, 2012)
34. S.J. Hawkes, Glass doesn't flow and doesn't crystallize and it isn't a liquid. J. Chem. Educ. **77**(7), 846 (2000)
35. E. Hempel, G. Hempel, A. Hensel, C. Schick, E. Donth, Characteristic length of dynamic glass transition near tg for a wide assortment of glass-forming substances. J. Phys. Chem. B **104**(11), 2460–2466 (2000)
36. J.M. Hutchinson, Physical aging of polymers. Progr. Polym. Sci. **20**(4), 703–760 (1995)
37. P. Jenniskens, D.F. Blake, Structural transitions in amorphous water ice and astrophysical implications. Science **265**(5173), 753–756 (1994)
38. B. Jerome, J. Commandeur, Dynamics of glasses below the glass transition. Nature **386**(6625), 589–592 (1997)
39. G.P. Johari, M. Goldstein, Molecular mobility in simple glasses. J. Phys. Chem. **74**(9), 2034–2035 (1970)
40. A.A. Jones, J.F. O'Gara, P. Inglefield, J.T. Bendler, A. Yee, K. Ngai, Proton spin relaxation and molecular motion in a bulk polycarbonate. Macromolecules **16**(4), 658–665 (1983)
41. J.L. Keddie, R.A. Jones, R.A. Cory, Size-dependent depression of the glass transition temperature in polymer films. Europhys. Lett. **27**(1), 59 (1994)
42. R. Kohlrausch, Nachtrag uber die elastiche nachwirkung beim cocon und glasladen. Ann. Phys. (Leipzig) **72**, 393 (1847)
43. J.S. Langer, Theories of glass formation and the glass transition. Rep. Progr. Phys. **77**(4), 042501 (2014)
44. L.-H. Lee, Adhesion of high polymers. I. Influence of diffusion, adsorption, and physical state on polymer adhesion. J. Polym. Sci. Part A-2 Polym. Phys. **5**(4), 751–760 (1967)
45. K. Li, P. Inglefield, A. Jones, J. Bendler, A. English, Heterogeneous microscopic mobility near the glass transition from NMR line shapes. Macromolecules **21**(10), 2940–2944 (1988)
46. Y. Liu, T. Russell, M. Samant, J. Stöhr, H. Brown, A. Cossy-Favre, J. Diaz, Surface relaxations in polymers. Macromolecules **30**(25), 7768–7771 (1997)
47. K.F. Mansfield, D.N. Theodorou, Molecular dynamics simulation of a glassy polymer surface. Macromolecules **24**(23), 6283–6294 (1991)
48. M.K. Mapes, S.F. Swallen, M. Ediger, Self-diffusion of supercooled o-terphenyl near the glass transition temperature. J. Phys. Chem. B **110**(1), 507–511 (2006)

49. A.M. Mayes, Glass transition of amorphous polymer surfaces. Macromolecules **27**(11), 3114–3115 (1994)
50. G.F. Meyers, B.M. DeKoven, J.T. Seitz, Is the molecular surface of polystyrene really glassy? Langmuir **8**(9), 2330–2335 (1992)
51. D. Prevosto, S. Capaccioli, K. Ngai, Origins of the two simultaneous mechanisms causing glass transition temperature reductions in high molecular weight freestanding polymer films. J. Chem. Phys. **140**(7), 074903 (2014)
52. A. Rachocki, E. Markiewicz, J. Tritt-Goc, Dielectric relaxation in cellulose and its derivatives. Acta Phys. Polon. Ser. A Gener. Phys. **108**(1), 137–146 (2005)
53. G. Reiter, Mobility of polymers in films thinner than their unperturbed size. Europhys. Lett. **23**(8), 579 (1993)
54. A. Richard et al., Interface and surface effects on the glass-transition temperature in thin polymer films. Farad. Discuss. **98**, 219–230 (1994)
55. R. Richert, K. Samwer, Enhanced diffusivity in supercooled liquids. N. J. Phys. **9**(2), 36 (2007)
56. E. Rössler, Indications for a change of diffusion mechanism in supercooled liquids. Phys. Rev. Lett. **65**(13), 1595 (1990)
57. S. Roy, C. Yue, Z. Wang, L. Anand, Thermal bonding of microfluidic devices: factors that affect interfacial strength of similar and dissimilar cyclic olefin copolymers. Sens. Actuat. B Chem. **161**(1), 1067–1073 (2012)
58. S.L. Shamblin, X. Tang, L. Chang, B.C. Hancock, M.J. Pikal, Characterization of the time scales of molecular motion in pharmaceutically important glasses. J. Phys. Chem. B **103**(20), 4113–4121 (1999)
59. R.S. Smith, B.D. Kay, Breaking through the glass ceiling: recent experimental approaches to probe the properties of supercooled liquids near the glass transition. J. Phys. Chem. Lett. **3**(6), 725–730 (2012)
60. F.H. Stillinger, A topographic view of supercooled liquids and glass formation. Science **267**(5206), 1935–1939 (1995)
61. F.H. Stillinger, A topographic view of supercooled liquids and glass formation. Science **267**(5206), 1935–1939 (1995)
62. S.F. Swallen, O. Urakawa, M. Mapes, M. Ediger, Self-diffusion and spatially heterogeneous dynamics in supercooled liquids near Tg, in *Slow Dynamics in Complex Systems: 3rd International Symposium on Slow Dynamics in Complex Systems*, vol. 708 (AIP Publishing, College Park, 2004), pp. 491–495
63. H. Takeuchi, R.-J. Roe, Molecular dynamics simulation of local chain motion in bulk amorphous polymers. I. dynamics above the glass transition. J. Chem. Phys. **94**(11), 7446–7457 (1991)
64. H. Takeuchi, R.-J. Roe, Molecular dynamics simulation of local chain motion in bulk amorphous polymers. II. Dynamics at glass transition. J. Chem. Phys. **94**(11), 7458–7465 (1991)
65. K. Tanaka, A. Taura, S.-R. Ge, A. Takahara, T. Kajiyama, Molecular weight dependence of surface dynamic viscoelastic properties for the monodisperse polystyrene film. Macromolecules **29**(8), 3040–3042 (1996)
66. G. Williams, D.C. Watts, Non-symmetrical dielectric relaxation behaviour arising from a simple empirical decay function. Trans. Farad. Soc. **66**, 80–85 (1970)
67. M.L. Williams, R.F. Landel, J.D. Ferry, The temperature dependence of relaxation mechanisms in amorphous polymers and other glass-forming liquids. J. Am. Chem. Soc. **77**(14), 3701–3707 (1955)
68. E.D. Zanotto, Do cathedral glasses flow? Am. J. Phys. **66**(5), 392–395 (1998)

Chapter 3
Physics of Deformation in Polymer Glasses and Deformation-Induced Molecular Mobility

Abstract Glassy polymers can exhibit a wide range of mechanical behaviors depending upon the rate of deformation, temperature, and the mechanical and thermal histories of the material. At the continuum level, their mechanical responses can be characterized as elastic, viscoelastic, and/or viscoplastic; *visco* indicating a time-dependent response. The macroscopic deformation of a glassy polymer is accompanied by molecular-level motions or rearrangements. Models of molecular motions during deformation have often been used in development of continuum-scale constitutive models. In this chapter, we provide a general overview of the mechanical behavior of glassy polymers (temperatures below the T_g) without elaborating on the formal framework of continuum mechanics. Discussions on the molecular mechanisms associated with the plastic deformation of glassy polymers and deformation-induced enhanced mobility are presented.

3.1 Introduction

Upon cooling, melts of thermoplastic polymers can crystallize or form an amorphous glass. In almost all cases, the microstructure obtained after cooling never fully crystallizes. Crystallization is essentially a kinetic process in which polymer chains organize at the molecular level to form an ordered crystalline structure. The degree of crystallinity (i.e., fraction of polymer present in crystalline form) depends upon the monomer type, interaction between monomers, chain rigidity, and cooling or quenching rates. High-chain rigidity, weaker monomer interactions, and fast cooling rates usually result in an amorphous state. Figure 3.1 shows a sample comparison of amorphous and semi-crystalline polymer microstructures. Polymer chains in an amorphous polymer are randomly coiled and there is no order in their spatial arrangement; whereas, in a semi-crystalline polymer the polymer chains form an ordered plate-like *crystalline-lamelle*. The lamelle are 10–20 nm in thickness, composed of partially aligned and folded molecular chains, and organize into structures called "spherulites" (Fig. 3.2).

Three common examples of amorphous glassy polymers are PMMA, PS, and PC. PMMA is a transparent thermoplastic used as a shatterproof replacement for

© The Author(s), under exclusive license to Springer Nature Switzerland AG 2021

N. Padhye, *Molecular Mobility in Deforming Polymer Glasses*,

SpringerBriefs in Materials, https://doi.org/10.1007/978-3-030-82559-1_3

Fig. 3.1 Crystalline vs amorphous domain microstructure

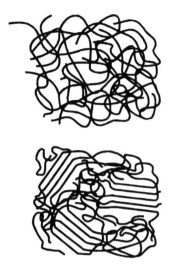

Fig. 3.2 Spherulites consisting of highly ordered lamellae

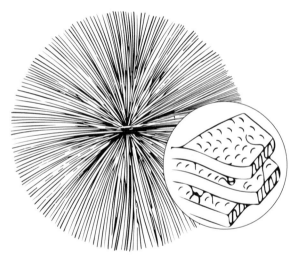

glass in car windows, windshields, television screens, laptops, and smartphone displays. Atactic PS is a hard, solid, and clear plastic used in foam materials, dining utensils, plastic cups, or other housewares, toys, covers and fixtures. PC is a temperature-resistant material with low light transmittance, used in safety glasses, bullet resistance, automotive door handles and headlights. Examples of widely used semi-crystalline polymers are polyethylene (PE) and polypropylene (PP). For semi-crystalline polymers, even if the crystalline domains are in a solid state (i.e., below the T_m), the amorphous domains can exist above their T_g or in a rubbery state. For example, high-density polyethylene (HDPE) has a T_g around -128 °C and T_m of 115–135 °C. Thus, at ambient temperatures, its amorphous phases are in a rubbery state while crystalline domains are below the T_m. The study of semi-crystalline polymers requires special attention because they essentially act as composite materials comprising crystalline and amorphous domains. In general, states of stresses or strains are not homogeneous across the crystalline and amorphous domains during polymer deformation.

Thermoplastics, both amorphous and semi-crystalline, are lightweight materials and can undergo appreciable permanent deformation. This property enables their usage as engineering materials for manufacturing of several end-products. The macroscopic mechanical behavior of thermoplastics is dependent on the conditions of operation, primarily the deformation rate and temperature, and the chemical composition and microstructural state prior to the deformation. The micromechanics of deformation of semi-crystalline and glassy polymers exhibit fundamental differences, which are often reflected in their gross mechanical behaviors at the continuum scale. A design engineer must pay attention to these aspects for structural or load-bearing applications to avoid undesirable failures. In this chapter, we will focus on the micromechanical aspects of deformation for glassy polymers and refer readers to [5, 9–11, 29, 30, 35, 48, 62, 67] (and references contained therein) for studies on semi-crystalline polymers. The microstructural aspects of deformation in rubber elasticity and viscoelastic polymeric liquids are also not discussed here. These topics are extensively dealt in the literature [22, 59].

The real challenge in modeling the mechanical behavior of a glassy (or for that matter any) polymer, is to satisfactorily derive and correlate the macroscopic physical-material behavior from the incurred molecular events. Molecular-level processes accompanied by the macroscopic deformation of a polymer are central to understanding and capturing the physics of polymer deformation. These molecular processes depend upon the internal state, chemistry, and processing conditions of the matter. Experimental techniques like small-angle X-ray scattering (SAXS), wide-angle X-ray scattering (WAXS), Raman spectroscopy, transmission electron microscopy (TEM), nuclear magnetic resonance (NMR), and small-angle neutron scattering (SANS), just to name a few, are being used in situ or post-deformation for studying the effects of deformation on microstructure and to draw inferences about the underlying mechanisms of relaxations at the microscopic level. Such understanding is critical for gaining better insights into the structure–property relationships and potentially tailoring the chemical composition or microstructure to ultimately control the macroscopic behavior for a specific need, or to develop mechanistically motivated continuum level models that incorporate the fundamental microscopic features of deformation.

In the next sections, we begin discussions on the general concepts of linear elasticity, viscoelasticity, and plasticity for glassy polymers. Phenomenological aspects of yielding, plastic flow, microstructural anisotropy, back-stress, and hardening are discussed, all with the considerations of large deformation. Brief discussions on the computational aspects of polymer modeling at the continuum scale are also provided. Standard exposition on the properties and mechanical behavior of the polymers can be found in [26, 33, 42, 57, 64].

3.2 Mechanical Behavior of Glassy Polymers

In general, glassy polymers can undergo large deformation and exhibit a vis-coplastic response (a rate-dependent behavior with plasticity). Uniaxial tension or compression, undertaken at different temperatures and strain rates, are standard methods to characterize the mechanical behavior of a polymer at the macroscopic scale. Before carrying out such mechanical characterization, it is advisable to investigate the nature of the polymer by assessing its amorphous or crystalline characteristics, degree of crystallinity, preparation history (including "ageing", which is a physical phenomenon of molecular-level relaxations in a glassy state that is out of thermodynamic equilibrium), and molecular and chemical details because these factors directly affect the mechanical behavior and hold the key for explaining the experimental observations.

Material characterization by mechanical testing essentially means obtaining characteristic mechanical data for "stress-strain" curves at different temperatures, deformation rates, or other test conditions. As such, this task appears to be straightforward, but care must be taken in understanding what type of a mechanical test should be performed, and corresponding interpretations of the experimental results. Ideally, one would like to obtain experimental data which provides the intrinsic constitutive material response; however, the specimen geometry, geometric or material imperfections, type of loading (state of the imposed deformation), and boundary conditions in a test can drastically effect the material response. These issues are not specific to polymers alone, but are relevant for other types of materials as well. Imposing a homogeneous state of deformation and measuring the corresponding material response during mechanical characterization is extremely important. Even simple tests like uniaxial tension or compression themselves have some challenges. In compression one must minimize (if not eliminate) the effects of the friction, and during a tension test geometric instability such as necking can occur and interfere with the "true" material behavior. Without delving into these details, we shall concentrate on the phenomenological features of a glassy polymer under the assumption that the material is subjected to *homogeneous* deformation, and this scenario would be consistent by imagining a uniaxial compression test without any frictional effects.

A schematic of a true stress-strain curve for a glassy polymer, under homoge-neous uniaxial compression, at a certain deformation rate, is shown in Fig. 3.3. The initial response is linear elastic, which is followed by a nonlinear elastic response. Depending upon the temperature and rate of deformation, the stress-strain curve in this linear and nonlinear elastic region may exhibit rate-sensitivity and, therefore, labeled with viscoelasticity. Although a sharp yield point is usually difficult to define for a glassy polymer, noticeable strains and geometric changes occur beyond the marked upper-yield point. Beyond this upper-yield point an interesting phenomenon of "strain softening", i.e., decrease in stress with increasing strain, is often noticed. The strain softening strongly depends on the preparation history of a glassy polymer, and the temperature at which the test is performed. Strain

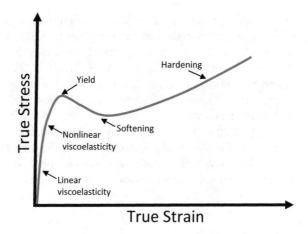

Fig. 3.3 Sample "stress-strain" curve for a glassy polymer

softening behavior is typically attributed to increase in local free volume at the microstructural scale, due to which resistance to plastic flow temporarily decreases until strain hardening is encountered. After the material has undergone a substantial amount of plastic deformation, the polymer chains gain preferred orientation at the molecular scale, and exhibit an increased resistance to further plastic flow (strain hardening). This regime is marked by large increase in the stress with the increasing strain. This behavior continues until fracture occurs. In case of glassy polymers the microstructure, comprising polymer chains, is randomly oriented, whereas in semi-crystalline polymers the deformation of crystalline domains, and the interdispersed amorphous regions, govern the mechanical behavior. References cited in the previous section have detailed discussions on theories and plausible mechanistic explanations for macroscopic mechanical behavior of glassy or semi-crystalline polymers. A detailed review of mechanistic viewpoints offered in the literature is beyond the current scope.

At the microscopic level, the rate-dependent behavior in glassy polymers occurs because molecular-level processes involving chain sliding, conformational changes, twisting or flexure of bonds during deformation require mechanical activation that depends on the rate of occurrences of these processes. If the timescales associated with deformation are larger than the relaxation times of the macromolecules or segments, then viscous (or time-dependent) behavior will be noted. Conversely, if the deformation rate (or temperature) is such that the polymer chains or segments have little or no opportunities to exhibit relaxations under imposed loadings, then an "elastic" response will be noted. At *small* strains, the mechanical response of glassy polymers can be described by the theory of linear elasticity or viscoelasticity. Linear elasticity idealizes the fully recoverable behavior, whereas viscoelasticity captures the rate-dependent behavior. At large strains, a complete viscoplastic description is required.

Temperature has a direct effect on the mechanical behavior of glassy polymers. At high temperatures, thermally activated rearrangements can occur readily and thereby aid the movement of polymer chains or segments. The rate-dependent behavior is also amplified as the temperature is increased. At temperatures well above the T_g, the barriers to segmental motions or conformational transitions are much lower compared with those encountered in the glassy state. For the same reason, macromolecules are often referred to be in a "locked-in" state below the T_g. High potential barriers, low kinetic energy, and less available free volume in a glassy state pose resistance at the microscopic level to conformational transitions, chain slippage, or other modes of relaxations during deformation in a glassy state. This is precisely the reason why irrecoverable plastic deformation requires cooperative rearrangements of a cluster of atoms (or molecules), and is mechanistically different from the macromolecular mobility in polymer melts.

3.2.1 Elasticity

Linear elasticity is strictly applicable if the material deformation is small and reversible. Even though, at the nanoscale, the microstructure of a glass is often heterogeneous and irreversible microscopic events can occur under small deformations, the bulk response of thermoplastics at the continuum scale, at small strains, can be idealized as recoverable and modeled through linear elasticity. Here, the symmetric Cauchy stress tensor (σ) is related to the symmetric small strain tensor (ϵ) constitutively through a fourth-order modulus tensor $\mathscr{C}$ as

$$\sigma = \mathscr{C} : \epsilon, \tag{3.1}$$

where the operator : indicates a double contraction of a fourth-order tensor ($\mathscr{C}$) with a second-order tensor (ϵ). The above relationship is also known as generalized Hooke's law. If the timescales for deformation of a glassy polymer are much smaller than the relaxation time of polymer chains/segments, temperatures are well below the T_g; viscous effects can be neglected all-together and an instantaneous response, under *small* strains, can be taken to be essentially reversible and elastic. At the molecular scale, small strains primarily involve stretching (or compression) against weak interatomic forces between the polymer atoms, or changes in bond angles through rotation of the polymer segments. In a linear-elastic regime, large-scale chain stretching along the polymer backbone is not triggered. Even at temperatures well below the T_g, the components of $\mathscr{C}$ are temperature-dependent (because they encode the aggregate molecular-level interactions during the relative motion of atoms or molecules). By imposing the symmetry of Cauchy stress and small strain tensor, $\mathscr{C}$ can have a maximum of 21 independent constants. In the simplest class of isotropic materials (where any arbitrary rotation of a material element renders its mechanical response unchanged), $\mathscr{C}$ comprises only two independent constants, commonly chosen as the elastic modulus E and Poisson's ratio ν. For an isotropic case, instead of E and ν, shear modulus μ and bulk modulus K are often chosen as the elastic constants because they quantify the shape- and volume-changing responses of the material, respectively. An isotropic elastic model of the material behavior holds true for unoriented glassy polymers because their microstructure comprises randomly organized polymer chains without preferential alignment of chains.

As mentioned earlier, the preparation history of a glass can affect its microstructure, and can render a process-induced microstructure. Under such circumstances a deviation from an isotropic behavior is expected. For example, PC, PMMA and PS are common examples of glassy polymers that show an isotropic response at small deformation (in the absence of a preformed or process-induced microstructure). Typically, a linear relationship between the stress and strain (e.g., in tensile or compression tests) can be noted up to strain levels of 5.0%. This is in major contrast with respect to the stress–strain behavior of metals, in which the linear relationship between stress and strain is valid up to a strain of only a fraction of 1%. The reason for this difference is because glassy polymers have a yield strength and elastic modulus that are up to two and three orders of magnitude, respectively, lower than those of metals. If strains exceed the yield limit, the behavior becomes more complex. By replacing the small strain tensor with the Hencky strain tensor the elastic response of a glassy polymer can be captured up to moderately large strains.

3.2.2   Viscoelasticity

Polymer glasses can be considered as highly viscous solids. One can think about the microscopic composition of a polymer as a combination of a viscous liquid and an elastic solid during deformation and, upon deformation, their microstructural dissipative processes depend on the rate of deformation. Thus, from an energy-balance perspective, the external work done on a viscoelastic polymer is not fully recoverable.

For glassy polymers, experiments have shown that if the temperatures and deformation rates are low, then the viscous effects are negligible; otherwise a viscoelastic behavior is observed, and viscoelasticity then refers to an elastic behavior in the presence of viscous effects. In viscoelasticity, the material deformation is usually recoverable and accompanied with internal losses. The viscoelastic behavior of a glassy polymer can be linear or nonlinear, and continues until the material undergoes yielding. Upon yielding, there is onset of plastic flow and permanent deformation. The post-yield behavior of the material can be rate-dependent, which is referred to as "viscoplasticity". In viscoplasticity, the permanent deformation processes exhibit a rate-dependent dissipation response. We will first discuss the classical ideas underlying viscoelasticity and then present the concepts of viscoplasticity in the next section.

Without getting into the details of the molecular motions associated with viscoelasticity (which are reserved for Sect. 3.3), let us consider two commonly encountered time-dependent behaviors under small deformation: creep and stress relaxation. Creep denotes the polymer deformation over long durations if an externally applied load is held constant. If an external load is applied suddenly, the initial material response is elastic; however, if the externally applied load is sustained over a long duration, then it causes a viscous flow in the material. In such situations, the total strain continues to increase. Conversely, if the material is subjected to a sudden external strain, which is held constant, then the initial material response will again be elastic (with the internal stresses related to the initial elastic strains through a linear elastic constitutive behavior). With the passage of time, there will be micro-structural rearrangements to cause trading of the elastic strains for the permanent deformation and thereby relaxing or diminishing the internal stresses. These behaviors can be understood through simple idealizations by using springs and dashpots. To convey these central ideas, we will restrict our discussions to one-dimensional deformation settings; however, generalizations to three-dimensional settings and large strains are necessary, and commonly employed to describe the mechanical behavior of polymer networks.

Consider a planar pure shear deformation, such that σ_{zz}, σ_{zx}, and σ_{zy} are equal to zero, and $\sigma_{yy} = -\sigma_{xx}$. For an ideal linear elastic behavior, the shear stress (σ_{xy}) is related to the shear strain (ϵ_{xy}) as $\sigma_{xy} = 2G\epsilon_{xy}$, where G is the shear modulus of elasticity. G is related to E and ν as $G = E/2(1 + \nu)$. In the case of an ideal linear viscous behavior (like that of a Newtonian fluid), under pure shear, the stress is related to strain rate as $\sigma_{xy} = 2\eta\dot{\epsilon}_{xy}$. To simplify our notation, we will drop the subscript xy, and the factor 2, and describe pure elastic and viscous behaviors by $\sigma = G\epsilon$ and $\sigma = \eta\dot{\epsilon}$, respectively. These linear elastic and viscous behaviors can be represented with a spring and dashpot, respectively. The purely elastic and viscous responses under a particular form of imposed stress and strain functions are shown in Figs. 3.4 and 3.5. For a purely elastic response, stress follows strain, and vice versa, and there is no "time lag" between the two. However, for a purely viscous material, a sudden application of stress does not cause any sudden strain increment; rather, if the stress is held constant, the strain increases linearly in accordance with

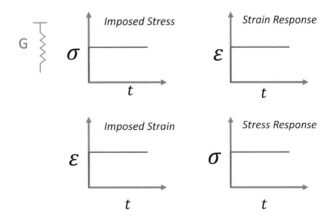

Fig. 3.4 Elastic response modeled by a spring. Stress and strain do not have a "time lag"

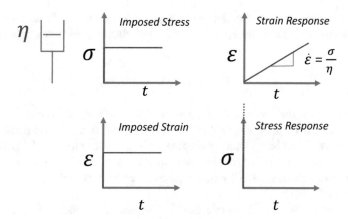

Fig. 3.5 Viscous response modeled by a dashpot. Stress and strain have a "time lag"

the viscous flow $\dot{\epsilon} = \sigma/\eta$. Conversely, if a sudden strain is imposed on a viscous liquid, implying an infinite strain rate, the instantaneous stress becomes unbounded at that instant; after which maintaining a constant strain over the time (i.e., zero strain rate) results in zero stress.

With these simple spring-and-dashpot elements, one can concoct elementary models for polymer behavior. Two such very basic mechanical models are: (a) the Maxwell model (a spring and a dashpot are connected in series), and (b) the Kelvin–Voigt model (a spring and a dashpot are connected in parallel). The Maxwell model is more appropriate to capture a stress-relaxation behavior, whereas the Kelvin-Voigt model is more suitable for reflect a creep response. These elementary models are not applicable for describing the general behaviors of real polymers due to several limitations, but a judicious combination of spring and dashpots can still be utilized to capture the salient phenomenological features of deformation in polymer glasses.

(a) **Maxwell model:** In this model, a spring and dashpot are connected in series so both the elements see identical stress at all times, but the total strain is the sum of individual strains:

$$\epsilon(t) = \epsilon_e(t) + \epsilon_v(t), \tag{3.2}$$

where $\epsilon(t)$ is the time-dependent total strain, and $\epsilon_e(t)$ and $\epsilon_v(t)$ are the time-dependent strains in elastic and viscous elements, respectively. Equation 3.2 can be transformed into the rate form as:

$$\dot{\epsilon}(t) = \dot{\epsilon}_e(t) + \dot{\epsilon}_v(t). \tag{3.3}$$

The strain rates in the above equations can be written in terms of time-dependent stress $\sigma(t)$ by using the elastic and viscous constitutive relationships as:

$$\dot{\epsilon}(t) = \dot{\sigma}(t)/G + \sigma(t)/\eta. \tag{3.4}$$

Taking the time derivative of the scalar stress (σ) is feasible because we have considered a 1D treatment, and there is "no rotation" of a material element. One could also formulate a three-dimensional viscoelastic response of a polymer if the strains are small, but rotations are finite. In such cases, the generalization of these simple models to 3D cases requires more care. We will not pursue those discussions here.

To study the behavior of this model under creep, consider an initial stress σ_o at $t = 0$, which is kept constant, i.e., $\sigma(t) = \sigma_o$. Because $\dot{\sigma}_o = 0$, according to Eq. 3.4, we get

$$\dot{\epsilon}(t) = \sigma_o/\eta, \tag{3.5}$$

and after integrating we have

$$\epsilon(t) = \epsilon_o + \sigma_o t/\eta, \tag{3.6}$$

where $\epsilon_o = \sigma/E$ is the initial elastic strain in the spring due to initially applied σ_o. Equation 3.6 reflects that, according to the Maxwell's model, during creep the strain increases continuously with time. Real polymers do not behave in this manner, in the sense that they achieve a final limiting strain at long times; which indicates a limitation of this model.

The stress relaxation behavior of the Maxwell model can also be analyzed by applying a constant strain ϵ_o at $t = 0$, and keeping it constant i.e. $\dot{\epsilon} = 0$. Under this deformation Eq. 3.4 becomes:

$$0 = \dot{\sigma}(t)/G + \sigma(t)/\eta. \tag{3.7}$$

If the suddenly applied strain ϵ_o causes a stress σ_o in the spring, then the solution to 3.7 is:

$$\sigma(t) = \sigma_o e^{-t/(\eta/G)}, \tag{3.8}$$

$\sigma_o = E\epsilon_o$ due to the initial elastic behavior. Overtime the elastic strains are traded for viscous strains, and stress decays to zero. In real glassy polymers, the stresses do not decay entirely to zero, which indicates a limitation of this model.

(b) **Kelvin-Voigt model:** This model also comprises a spring and dashpot, and they are connected in parallel. Due to this arrangement of the elements, they see identical strain but the total stress on the assembly is the sum of the individual element stresses. This can be expressed as:

$$\sigma(t) = \sigma_e(t) + \sigma_v(t), \tag{3.9}$$

$\sigma(t)$, $\sigma_e(t)$, and $\sigma_v(t)$ are time-dependent total stress, spring stress, and dashpot stress, respectively. We can now re-write Eq. 3.9 as

$$\sigma(t) = G\epsilon(t) + \eta\dot{\epsilon}(t), \tag{3.10}$$

where we have used the constitutive relationship, and the fact that elastic and viscous strains (or strain rates) are identical in both elements. For studying the creep behavior of the Kelvin-Voigt model, we subject it to a constant stress $\sigma(t) = \sigma_o$ at $t = 0$. Equation 3.10 then becomes

$$\sigma_o = G\epsilon(t) + \eta\dot{\epsilon}(t). \tag{3.11}$$

The solution of Eq. 3.11, strain as a function of time, is

$$\epsilon(t) = \sigma_o/G[1 - e^{-t/(\eta/G)}]. \tag{3.12}$$

According to Eq. 3.12, the instantaneous response by applying load σ_o at $t = 0$ results in zero strain. This denotes that the strains cannot grow instantaneously in the spring because there is a viscous dashpot attached in parallel. The strains in the viscous dashpot are initially zero, and develop over time by integrating the strain rate. At large times ($t \to \infty$), the strains reach their elastic limit σ_o/G. Although this model captures a limiting feature of the behavior of real glassy polymer—at large times stress reaches an asymptotic value—it does not capture the initial instantaneous elastic response.

In the case of stress relaxation, if we apply a sudden strain ϵ_o and keep the strain constant with $\epsilon(t) = \epsilon_o$, then there is an initial finite elastic response in the spring and unbounded growth of stress in the dashpot. The unbounded stress in the dashpot immediately goes to zero as the strain is held constant, and the elastic strain in the spring is constant, which also keeps the stress in the spring constant. Thus, there is no mechanism of stress relaxation in this model. It is quite clear that these models cannot describe real behavior. Creep is the continuous accumulation of strain under an externally applied load. A sustained applied load is a concern for loaded structures that are required to maintain their geometry without extensive deformation (e.g., medical sutures, or implants). Stress relaxation occurs if a structure has a fixed configuration resulting in an initial imposed stress (e.g., press-fit component), where the initial stress relaxes over time and may result in the assembly coming apart. Either of these two phenomena can result in an adverse outcome in an engineered system and, therefore, must be anticipated and analyzed in viscoelastic materials.

Fig. 3.6 Time- and temperature-dependent shear modulus

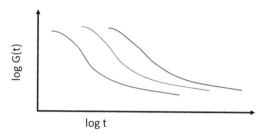

Time Temperature Equivalence Superposition So far we have discussed the time-dependent viscoelastic behavior of a glassy polymer, but temperature plays an equally important part. The time-dependent shear modulus $G(t)$ can better be represented as $G(t, T)$, thereby accounting for temperature dependence. A schematic of temperature dependence of the shear modulus is shown in Fig. 3.6. We observe that, as the temperature increases, the time-dependent shear modulus decreases.

Experiments have shown that the time- and temperature-dependent behavior of glassy polymers can be consolidated into a single master curve. Consider a reference temperature T^* at which one has obtained the time-dependent shear modulus $G(t, T^*)$. The time-dependence of $G(t, T^*)$ reflects the molecular-level processes, at T^*, that occur over time to cause deformation. A material subjected to pure shear stress σ, at T^*, will incur a time-dependent shear strain $\epsilon(t, T^*) = \sigma/G(t, T^*)$, because the shear modulus $G(t, T^*)$ changes with time. The accumulated strain $\epsilon(t, T^*)$ at T^* increases with time because $G(t, T^*)$ decreases with time. Let us call T^* the "reference temperature". Suppose we change the temperature to a higher value at T, and now the molecular-level processes at T are accelerated by a factor a_T compared with those at T^*. This would imply that, to achieve a same level of strain (or deformation) at T, the time needed would be a factor a_T less than the time required at T^*. Alternatively, the strain (or deformation) noted at time $a_T t$ and temperature T^* would be same as that noted at time t and temperature T. We can express this as:

$$\epsilon(t, T) = \epsilon(a_T t, T^*). \tag{3.13}$$

Equation 3.13 can be represented in terms of stress and time-dependent modulus as:

$$\sigma/G(t, T) = \sigma/G(a_T t, T^*), \tag{3.14}$$

implying that $G(t, T) = G(a_T t, T^*)$, which is the time temperature superposition principle. Experimentally measured moduli over different times and temperatures have been found to follow this behavior. Obviously, the factor a_T depends on T itself. Polymer scientists have tried to explain this behavior by noting that the relaxation time $\tau(T)$ at temperature T is lowered by a factor a_T compared with the relaxation time $\tau(T^*)$, i.e.:

$$\tau(T) = \frac{\tau(T^*)}{a_T} \tag{3.15}$$

or,

$$a_T = \frac{\tau(T^*)}{\tau(T)}. \tag{3.16}$$

Here, we have implicitly assumed that the polymer deforms via a single relaxation mechanism with relaxation time τ. We emphasize that these models are not applicable at large strains. Also, in reality there is a spectrum of relaxation mechanisms with a distribution of relaxation times. The temperature dependence of a_T is commonly modeled by Williams-Landel-Ferry (WLF) equation:

$$log(a_T) = \frac{-C_1(T - T^*)}{C_2 + (T - T^*)}, \tag{3.17}$$

where T^* is some reference temperature, and C_1 and C_2 are empirical constants. T^* is often chosen as T_g. The derivation of the above equation is available in standard textbooks referred earlier.

3.2.3 Plastic Deformation

As stated previously, plastic (or permanent) deformation in polymers is noted beyond the yield point on a stress-strain curve. The yield point in glassy polymers is not unique; it depends on the temperature and rate of deformation, as well as the thermal and preparation history of the polymer. A polymer glass is not in a thermodynamic equilibrium, so slow structural relaxations continue to take place in the glassy state to lower the internal free energy. This process is known as "ageing", and is also accompanied by lowering of the free volume. Ageing can alter the mechanical response of the polymer drastically. We will not discuss ageing further here and focus rather on general phenomenological aspects of plastic deformation. Higher temperatures and lower strain-rates lower the yield stress of a glassy polymer. Below the T_g, upon reaching the yield point, first a strain-softening behavior is observed (i.e., the flow stress required to cause plastic flow decreases with increasing strain). Then, upon further plastic deformation strain hardening is observed.

There is a continuous evolution of microstructure during plastic deformation. The uniaxial stress-strain curves for a glassy polymer are characteristically different depending on whether the sample is loaded in tension or compression. Yield strengths obtained from compression tests are usually greater than those from tensile tests. The compression tests are accompanied by a negative (compressive) mean normal stress, which have been postulated to increase the yield strength whereas, during tension a dilatational (volume increasing) mean normal stress is believed to reduce the yield strength. Strain-softening is observed in the post-yield regime in compression and tension, and is associated with the onset of many spatial microstructural relaxation events and localization of strains into narrow band-like regions that sustain deformation with decreasing stress. The strain-softening saturates with increments of plastic strain and eventually gives way to strain-induced hardening within the polymer. At large plastic strains (roughly greater than 30%) the underlying glassy network of the polymer begins to deform as a whole and develops noticeable anisotropy, and molecular-chain or segmental reorientations correlate with the directions of principal stretches. During strain-hardening, the mechanical stresses required to continue plastic deformation rise rapidly. A key distinctive feature of tensile deformation in glassy polymers during plastic deformation (where the mean normal stress and maximum principal stress are positive) is the development of "crazes" or the phenomenon of "crazing". "Crazes" are defined as regions of micro-voids within a polymer comprising polymer fibrillates, and exhibiting low density. PS and PMMA are common examples of glassy polymers that show crazing. Conditions favoring the appearance of crazes are positive mean normal stress (dilatational) and a positive (tensile) maximum principal stress. Both of these factors enable the "loosening" of the polymer entanglements and facilitate crazing. If the conditions for crazing are met, then the formation and growth of crazes continues with the lowering of local material density, development of chain orientation and, ultimately, chain scission. Breakdown in the craze matter causes the growth of cracks and, ultimately, material separation from fracture. For example, see [4, 24, 25] and the references cited within them for more details.

The plastic flow in glassy polymers is rate- and temperature-dependent, and usually described by specification of a viscoplastic flow rule that captures these dependencies along with other phenomenological features (that have been discussed). Several 3D continuum scale models, isothermal or thermomechanically coupled, describing the behavior of glassy polymers at large deformations have been proposed in [1–3, 12, 13, 51, 55]. Atomistic or molecular dynamics simulations have also been deployed to reveal the underlying plausible mechanisms and microstructural changes during plastic deformation (e.g., [6, 28, 30, 43, 52, 58]). In no way are these an exhaustive list of references, and the reader is encouraged to study various viewpoints and intricacies of the subject beyond the ones presented here.

Lastly, the classical rate-independent plasticity models, characterized by a yield criterion and rate-independent flow rule, can be utilized for glassy polymers if the rates of deformation are slow and temperatures are well-below the T_g. The Mises yield criterion and normality flow rule can capture the plasticity behavior of glassy-polymers up to moderately large levels of deformation prior to development of appreciable anisotropy (after which an anisotropic yield criterion would be needed). Such an approach is computationally more efficient for the simulation purposes, for example, a rolling process modeled with rate-independent plasticity [47]. The prime reason for this recommendation is that the rules for viscoplastic flow (and associated time-dependent material parameters with their specified time evolution) are computationally quite expensive for numerical time-integration in a finite element procedure. If the rate-effects are known *a priori* to be negligible then a rate-independent material model is computationally cheaper to adopt during an incremental finite element procedure.

3.3   Molecular Picture of Deformation and Enhanced Mobility

So far, we have mostly discussed the phenomenological features of deformation in glassy polymers. Microstructural changes accompanying the deformation are critical for gaining fundamental understanding of the material mechanical behavior. In this section, we will discuss the notion of deformation-induced molecular mobility, and present some central ideas underlying the mechanism of mobility of polymer chains and segments during deformation.

Unlike molecular liquids, glassy polymers well below their T_g are in a solid state and macromolecules are in a kinetically trapped state, so long-range diffusive motion cannot be demonstrated. The solid state of polymers can be regarded as a highly viscous liquid, wherein the activity of long-chain molecules is highly suppressed, and on experimental timescales macromolecules are effectively frozen. This is a universally accepted nature of the "glassy state". In 1936, Eyring [21] presented a theory for stress-induced molecular mobility in glasses using an absolute reaction-rate approach to demonstrate the effect of stress on lowering of viscosity in certain gels, glasses or crystals and its effect on the enhancement of molecular mobility. This model is schematically shown in Fig. 3.7. According to Erying's model:

$$\frac{1}{\nu_1} = \tau_\alpha(T, \sigma) = \tau_o exp\left[\frac{E_A - \sigma V_a^*}{K_B T}\right]. \tag{3.18}$$

Here, ν_1 is the jump frequency (inverse of the relaxation time τ_α) of the molecules in the direction of applied shear stress σ, τ_o is a constant (of the order of atomic vibration timescale), V_a^* is the activated volume, E_A is the potential barrier that molecules must overcome in going from one configuration to another. Equation 3.18 states that the relaxation time decreases for processes in which the direction of the applied shear stresses lowers the barrier, so that molecular transport is facilitated in the direction of shear stresses. This is also referred to as "stress activation". Eyring employed this model to explain the mechanical behavior of rubbery solids, including behaviors such as creep, relaxation, and hysteresis. However, this model is applicable only in the regimes of linear viscoelasticity for glassy polymers. The temperature and strain-rate dependence of the yield stress is also adequately

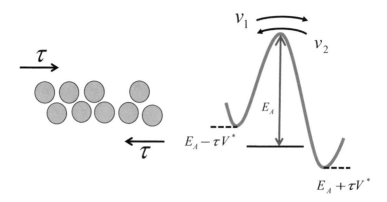

Fig. 3.7 Eyring's model of mechanically-induced molecular mobility. This conceptual picture also enables us to think about the plastic deformation in glassy polymers as a thermally activated process

described by it (see for example [63]), but it fails to capture the dramatic changes in the energy landscapes associated with plastic deformation in a glass.

The plastic deformation of glassy polymers at a continuum scale is understood in terms of localized step-like shear cooperative displacements of lengthy chain segments, and the unit plastic rearrangements are known as "shear transformations" [5]. According to molecular dynamics simulations [56], slippage of chains is the underlying feature of a shear transformation. Shear transformations are events of spatial rearrangements of molecular clusters causing stress relaxation. Consider the scenario shown in Fig. 3.8. Well below the T_g, the polymer chains are kinetically trapped in their local configurations, and the timescales for mobility (specifically, translation motions) of these chains are extremely large. However, application of shear stress on the material element causes its deformation, and the polymer chain under consideration changes its orientation: first elastically, and then, due to some local perturbation, it relaxes plastically while overcoming the potential barrier set up due to neighboring molecules.

The irreversible events associated with shear transformations directly contribute to the accelerated mobility of polymers during deformation. If no stresses or deformation were applied and the temperature was held far below the T_g, then the transition of the mean configuration of a polymer chain from its kinetically trapped configuration would not occur on experimental timescales. However, qualitatively speaking, application of the *stress enhances the mobility* of the polymer molecules as it relaxes, and changes their configurations on experimental timescales. The effects of such cumulative events during active plastic deformation characterize the enhanced dynamics in deforming glasses below their T_g. Because of the kinetically-trapped state of a glass, it is implied that any cooperative segmental relaxations or long-range diffusive motions of chains are severely restricted, secondary relaxation processes (those corresponding to vibrations of side groups such as β, γ, δ, etc.

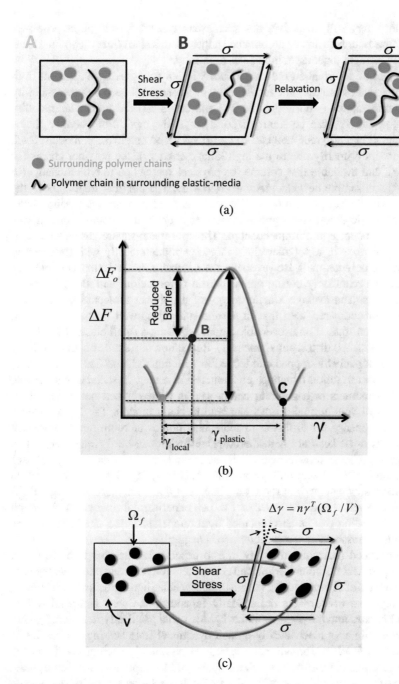

Fig. 3.8 Mechanism of plastic deformation and shear transformation in glassy polymers [5]. (**a**) A unit shear transformation in a kinetically trapped state under shear stress comprising an initial elastic shear strain followed by plastic-relaxation of a polymer chain. (**b**) Free energy landscape of a polymer chain during a unit shear transformation. (**c**) Accumulation of several shear transformations leads to macroscopic plastic deformation

relaxations) may still be active, but such weak secondary relaxation processes cannot cause interdiffusion or pronounced adhesion in a short time if two interfaces are simply brought together in molecular proximity.

The stress of deformation-induced mobility in polymer glasses is different from thermally activated motion. The fundamental differences between polymer mobility at high temperatures (well above the T_g) and stress-assisted molecular mobility (well below the T_g) can be summarized. That is, the motion of polymer chains (or segments) in a polymer melt can be described based on diffusion models. Such motion occurs primarily due to the high kinetic energy of the polymer chains (or segments), and available free volume (or physical space) due to which chains (or segments) can sample new orientations effectively. The polymer melts (above the T_g) are spatially homogeneous and thermodynamically in an equilibrium state. However, plastic deformation and the associated enhanced mobility in a glassy polymer is not an equilibrium concept. The root-mean-square displacement of the center of mass of a polymer chain increases monotonically with time during diffusion in a polymer melt. However, the mechanically assisted enhanced mobility in polymers occurs only during active plastic deformation, and stops if plastic straining stops. The average kinetic energy of a polymer molecule is large in a polymer melt compared with that in a solid-state glass well below its T_g. In a glassy or frozen state, if an irreversible step must occur, a small cluster of polymer chains (or segments) distort simultaneously. Relaxation of a single chain is unlikely because it is kinetically trapped and locked in with other chains. This relaxation is different from relaxation dynamics encountered in a melt (associated with Rouse or reptation motions of a polymer chain, say, in a crosslinked network), where sufficient local volume fluctuations can lead to chain motion. The typical values of activation energy for diffusion in molecular liquids at room temperature can be as low as 5–10 kcal/mol and, therefore, diffusion can be thermally activated. However, for a shear transformation, the activation energy in the case of glasses can be up to 350–400 kcal/mol (approximately 15 ev) [5] so plastic deformation cannot occur spontaneously at room temperatures (where the thermal energy, $K_B T$, at room temperature is of the order of 1 ev) on experimental timescales. Although plastic deformation can be accompanied by a temperature rise, at relatively slow strain rates, the associated temperature rise is negligible.

The enhanced molecular mobility due to deformation has been confirmed in several experimental studies: Argon and co-workers [68] demonstrated that the case II sorption rates of low-molecular weight diluent species into a plastically deforming glassy poly(ether-imide) were dramatically enhanced, and were comparable with the sorption rates into the polymer at the T_g. Also, that plastically-deforming glassy polymers exhibited a mechanically dilated dynamical state bearing similarities to the molecular-level conformational rearrangements taking place at the T_g in the absence of active deformation. A related study [8] also reported an increase in the case II front velocity (of approximately 6.5 times) when an out-of-surface tensile stress was applied. In this study, it is likely that applied tensile loading normal to the diffusion front caused molecular rearrangements of polymer chains or segments that enabled faster movement of the liquid penetrant. Lee et al. [38] showed that uniaxial deformation of PMMA at 19 K below its T_g exhibited an increased molecular mobility up to 1000 times. The authors used an optical photobleaching technique to quantitatively characterize the changes in molecular mobility during the active deformation in uniaxial tensile creep. It should be noted that the authors actually inferred the active mobility of polymer chains or segments based on the accelerated dynamics of reorientation of the photobleaching agent during deformation, and this mobility was characterized in terms of reduction in the relaxation time of the reorientation of a dilute molecular probe. See [36, 37] (and the cited references therein) on dye reorientation as a probe for stress-induced mobility in polymer glasses. Lee et al. [39] studied the stress relaxation response of glassy PMMA just below its T_g by applying constant strain-rate uniaxial tensile deformation at various locations on the stress strain curve, including the yield and post-yield region. They analyzed the macroscopic mobility from the analysis of the relaxation response, and found a decrease in the relaxation time up to a factor of 3, with the fastest relaxation occurring in the post-yield softening region. In [31, 32], similar ideas were presented, for example, to deploy stress relaxation experiments to provide indirect evidence of enhanced molecular mobility during deformation. In [40], the authors used NMR on deuterated semi-crystalline Nylon 6. They reported enhanced conformational dynamics in the amorphous regions of Nylon when deformation was carried out near the T_g. Molecular dynamics simulations have also revealed enhanced molecular mobility during active deformation of a glass. For example, using a United Atom (UA) model for glassy polyethylene at 200 K, Capaldi et al. [15, 16] showed that dihedral transition rates increased by nearly one order of magnitude as the bulk material underwent yield and plastic flow under compression. Such transitions were correlated along the length of the chain; remarkably, no correlation of transition rate with local density was observed. Furthermore, these dihedral transition rates were accelerated only under active deformation; during cyclic deformation, the rates returned to their quiescent values during periods of constant (but nonzero) strain. Other examples of mechanically activated molecular mobility can be found in [17–19, 23, 34, 44–46, 49, 50, 53, 65, 66]. An interesting mechanism of toughening during deformation (and associated mobility) has been reported for thermoplastics (which typically undergo crazing and demonstrate a relatively brittle behavior in tension): in [7], the authors proposed blending of a brittle (homo)PS with a very small concentrations of low-molecular weight elastomeric components, and claimed very effective toughening through a process of solvent crazing. By incorporating a small amount (low percentage by weight) of low-molecular weight polybutadiene (PB) rubber into the PS matrix, lowering in the craze yield strength (enabling large dissipation during crack propagation and, thus, enhancement in the toughness) was reported. The noted toughening was attributed to the craze matter intersecting with the pools of dissolved plasticizing agent (PB) and subsequent distribution of the plasticizer into the craze region and surfaces during deformation, thereby mitigating brittle crack-propagation. The deformation-induced mobility of polymer chains or segments likely enabled the dispersion of the plasticizing agent from the pools of dissolved PB into the craze matter, thus enhancing the ductility. By utilizing the concept of enhanced molecular mobility of chains (or segments) during deformation of polymer glasses, recently a new phenomenon of sub-T_g, solid-state, plasticity-induced bonding was reported wherein solid-state glassy polymers were bonded through direct application of plastic deformation. Since thermoplastics form an important class of engineering materials, and several everyday commercial products made out of plastics by bonding of plastics, so the next chapter is dedicated to this new class of polymer bonding.

It is clear that enhanced polymer mobility is intrinsically connected to irreversible deformation, and plays an important part in determining the macroscopic mechanical properties of glassy polymers. The dynamical processes that accompany plastic deformation ultimately render a microstructure, and the resulting properties, both, during active deformation and post-deformation are thus tied to the mobility of polymer chains and segments. Understanding, characterizing, or modeling the polymer mobility during deformation, and ultimately controlling them (through tailoring at the molecular level or blending with additives or otherwise) can lead to preferential material behaviors. However, a comprehensive characterization of the effect of molecular motions on the mechanical behavior has not been fully achieved. Even after several decades of experimental work and computer simulations, there are no deterministic means of revealing molecular mechanisms during large strain plastic deformation in polymers that can explain macroscopic findings with complete certainty. In fact, on several occasions, conflicting and multiple viewpoints have emerged. Advances in polymer characterization techniques—at wide range of length- and timescales through analytical tools with all the variants of microscopy, spectroscopy, scattering, and relaxation–have enabled the study of polymer microstructure, but such techniques have been unable to provide direct, deterministic, and continuous time tracking for all polymer chains, molecules, or functional groups, in the solid state during deformation. NMR has emerged as a powerful tool to study the conformational distribution in glassy polymers upon deformation, for example, see [14, 20, 27, 60, 61]. However, the definitive details on the dynamical molecular processes that lead to the inferred microstructures have not been provided. The strengths and limitations of NMR and other spectroscopy techniques for solid-state polymers are described in [41, 54]. Despite such limitations, these techniques will continue to advance our understanding of deformation in polymers at the molecular scale. There are also major challenges in deriving exhaustive information about mechanisms during deformation of polymer glasses through molecular simulations. The main source of difficulty is the computational infeasibility of simulating detailed molecular-level models of sufficiently large systems, representative of the continuum-scale behavior over a range of experimental timescales. Currently, even with the most powerful available computing platforms, exact simulations of the continuum-scale behavior of glassy polymers (and their composites) over experimental timescales is not feasible. It is expected, however, that developments in high-performance computing for molecular simulations will continue to enable better understanding of induced molecular mobility due to active deformation, and its ultimate effect on the mechanical behavior. Importantly, even if the simulation capabilities reached a level such that the continuum-scale mechanical behavior of polymers could be predicted quantitatively from molecular simulations alone, direct experimental means would be required to fully verify the accuracy of molecular-level processes (revealed through simulations) that conform with the observed mechanical behaviors. Until we reach such a stage, a judicious and iterative multiscale modeling strategy that carefully and accurately links the material behavior at different length scales appears to be a promising approach.

References

1. N.M. Ames, V. Srivastava, S.A. Chester, L. Anand, A thermo-mechanically coupled theory for large deformations of amorphous polymers. part II: applications. Int. J. Plast. **25**(8), 1495–1539 (2009)
2. L. Anand, M.E. Gurtin, A theory of amorphous solids undergoing large deformations, with application to polymeric glasses. Int. J. Solids Struct. **40**(6), 1465–1487 (2003)
3. L. Anand, N.M. Ames, V. Srivastava, S.A. Chester, A thermo-mechanically coupled theory for large deformations of amorphous polymers. Part I: formulation. Int. J. Plast. **25**(8), 1474–1494 (2009)
4. A. Argon, Craze initiation in glassy polymers—revisited. Polymer **52**(10), 2319–2327 (2011)
5. A.S. Argon, *The Physics of Deformation and Fracture of Polymers*. Cambridge University Press, Cambridge (2013)
6. A.S. Argon, M. Hutnik, P.H. Mott, U.W. Suter, Simulation of inelastic deformation in glassy polypropylene and polycarbonate. Technical report, Massachusetts Inst of Tech, Cambridge, 1989
7. A.S. Argon, R. Cohen, O. Gebizlioglu, H. Brown, E. Kramer, A new mechanism of toughening glassy polymers. 2. Theoretical approach. Macromolecules **23**(17), 3975–3982 (1990)
8. A. Argon, R. Cohen, A. Patel, A mechanistic model of case II diffusion of a diluent into a glassy polymer. Polymer **40**(25), 6991–7012 (1999)
9. A. Argon, A. Galeski, T. Kazmierczak, Rate mechanisms of plasticity in semi-crystalline polyethylene. Polymer **46**(25), 11798–11805 (2005)
10. S. Balijepalli, G. Rutledge, Simulation study of semi-crystalline polymer interphases, in *Macromolecular Symposia*, vol. 133 (Wiley Online Library, New York, 1998), pp. 71–99
11. Z. Bartczak, A. Galeski, Plasticity of semicrystalline polymers, in *Macromolecular Symposia*, vol. 294 (Wiley Online Library, New York, 2010), pp. 67–90
12. J.S. Bergstrom, *Mechanics of Solid Polymers: Theory and Computational Modeling* (William Andrew, Amsterdam, 2015)
13. M.C. Boyce, D.M. Parks, A.S. Argon, Large inelastic deformation of glassy polymers. Part I: rate dependent constitutive model. Mech. Mater. **7**(1), 15–33 (1988)
14. A.J. Brandolini, D.D. Hills, *NMR Spectra of Polymers and Polymer Additives* (CRC Press, Boca Raton, FL, 2000)
15. F.M. Capaldi, M.C. Boyce, G.C. Rutledge, Enhanced mobility accompanies the active deformation of a glassy amorphous polymer. Phys. Rev. Lett. **89**(17), 175505 (2002)
16. F.M. Capaldi, M.C. Boyce, G.C. Rutledge, Molecular response of a glassy polymer to active deformation. Polymer **45**(4), 1391–1399 (2004)
17. K. Chen, K.S. Schweizer, Stress-enhanced mobility and dynamic yielding in polymer glasses. Europhys. Lett. **79**(2), 26006 (2007)
18. Y.G. Chung, D.J. Lacks, Atomic mobility in strained glassy polymers: the role of fold catastrophes on the potential energy surface. J. Polym. Sci. Part B Polym. Phys. **50**(24), 1733–1739 (2012)
19. Y.G. Chung, D.J. Lacks, How deformation enhances mobility in a polymer glass. Macromolecules **45**(10), 4416–4421 (2012)

20. E. Egorov, V. Zhizhenkov, NMR studies of molecular mobility in uniaxially stretched oriented polymers. J. Polym. Sci. Polym. Phys. Ed. **20**(7), 1089–1106 (1982)
21. H. Eyring, Viscosity, plasticity, and diffusion as examples of absolute reaction rates. J. Chem. Phys. **4**(4), 283–291 (2004)
22. J.D. Ferry, J.D. Ferry, *Viscoelastic Properties of Polymers* (Wiley, New York, 1980)
23. B. Frank, A.P. Gast, T.P. Russell, H.R. Brown, C. Hawker, Polymer mobility in thin films. Macromolecules **29**(20), 6531–6534 (1996)
24. B. Gearing, L. Anand, Notch-sensitive fracture of polycarbonate. Int. J. Solids Struct. **41**(3–4), 827–845 (2004)
25. B. Gearing, L. Anand, On modeling the deformation and fracture response of glassy polymers due to shear-yielding and crazing. Int. J. Solids Struct. **41**(11–12), 3125–3150 (2004)
26. R.N. Haward, *The Physics of Glassy Polymers* (Springer Science & Business Media, Dordrecht, 2012)
27. C. Hedesiu, D.E. Demco, K. Remerie, B. Blümich, V.M. Litvinov, Study of uniaxially stretched isotactic poly (propylene) by 1h solid-state NMR and IR spectroscopy. Macromol. Chem. Phys. **209**(7), 734–745 (2008)
28. R.S. Hoy, M.O. Robbins, Strain hardening of polymer glasses: effect of entanglement density, temperature, and rate. J. Polym. Sci. Part B Polym. Phys. **44**(24), 3487–3500 (2006)
29. S. Jabbari-Farouji, O. Lame, M. Perez, J. Rottler, J.-L. Barrat, Role of the intercrystalline tie chains network in the mechanical response of semicrystalline polymers. Phys. Rev. Lett. **118**(21), 217802 (2017)
30. S. Jabbari-Farouji, J. Rottler, O. Lame, A. Makke, M. Perez, J.-L. Barrat, Plastic deformation mechanisms of semicrystalline and amorphous polymers. ACS Macro Lett. **4**(2), 147–150 (2015)
31. J.W. Kim, G.A. Medvedev, J.M. Caruthers, Mobility evolution during tri-axial deformation of a glassy polymer. Polymer **55**(6), 1570–1573 (2014)
32. J.M. Kropka, K.N. Long, Can stress relaxation experiments be used to assess deformation induced mobility in glassy polymers? Technical report, Sandia National Lab (SNL-NM), Albuquerque, NM (United States), 2016
33. R.F. Landel, L.E. Nielsen, *Mechanical Properties of Polymers and Composites* (CRC Press, Boca Raton, 1993)
34. H.-N. Lee, M. Ediger, Mechanical rejuvenation in poly (methyl methacrylate) glasses? Molecular mobility after deformation. Macromolecules **43**(13), 5863–5873 (2010)
35. B.J. Lee, A.S. Argon, D.M. Parks, S. Ahzi, Z. Bartczak, Simulation of large strain plastic deformation and texture evolution in high density polyethylene. Polymer **34**(17), 3555–3575 (1993)
36. H.-N. Lee, K. Paeng, S. Swallen, M. Ediger, Dye reorientation as a probe of stress-induced mobility in pmma glasses, in *APS Meeting Abstracts*, vol. 1 (2008), pp. 18010
37. H.-N. Lee, K. Paeng, S.F. Swallen, M. Ediger, Dye reorientation as a probe of stress-induced mobility in polymer glasses. J. Chem. Phys. **128**(13), 134902 (2008)
38. H.-N. Lee, K. Paeng, S.F. Swallen, M.D. Ediger, Direct measurement of molecular mobility in actively deformed polymer glasses. Science **323**(5911), 231–234 (2009)
39. E.-W. Lee, G.A. Medvedev, J.M. Caruthers, Deformation induced evolution of mobility in PMMA. J. Polym. Sci. Part B: Polym. Phys. **48**(22), 2399–2401 (2010)
40. L.S. Loo, R.E. Cohen, K.K. Gleason, Chain mobility in the amorphous region of nylon 6 observed under active uniaxial deformation. Science **288**(5463), 116–119 (2000)
41. A. Martínez-Richa, R.L. Silvestri, *Developments in Solid-State NMR Spectroscopy of Polymer Systems* (IntechOpen, London, 2017)
42. G.H. Michler, F.J. Baltá-Calleja, *Mechanical Properties of Polymers Based on Nanostructure and Morphology* (CRC Press, Boca Raton, 2016)
43. P. Mott, A. Argon, U. Suter, Atomistic modelling of plastic deformation of glassy polymers. Philos. Mag. A **67**(4), 931–978 (1993)
44. E. Munch, J.-M. Pelletier, B. Sixou, G. Vigier, Characterization of the drastic increase in molecular mobility of a deformed amorphous polymer. Phys. Rev. Lett. **97**(20), 207801 (2006)

45. E. Munch, J.-M. Pelletier, G. Vigier, Increase in molecular mobility of an amorphous polymer deformed below TG. J. Polym. Sci. Part B Polym. Phys. **46**(5), 497–505 (2008)
46. E. Oleynik, Plastic deformation and mobility in glassy polymers, in *Relaxation in Polymers* (Springer, New York, 1989), pp. 140–150
47. N. Padhye, Sub-Tg, solid-state, plasticity-induced bonding of polymeric films and continuous forming. PhD thesis, Massachusetts Institute of Technology, 2015
48. D.M. Parks, S. Ahzi, Polycrystalline plastic deformation and texture evolution for crystals lacking five independent slip systems. J. Mech. Phys. Solids **38**(5), 701–724 (1990)
49. R.A. Riggleman, H.-N. Lee, M. Ediger, J.J. De Pablo, Heterogeneous dynamics during deformation of a polymer glass. Soft Matter **6**(2), 287–291 (2010)
50. R.A. Riggleman, K.S. Schweizer, J.J.D. Pablo, Nonlinear creep in a polymer glass. Macromolecules **41**(13), 4969–4977 (2008)
51. C.B. Roth, *Polymer Glasses* (CRC Press, Boca Raton, 2016)
52. J. Rottler, S. Barsky, M.O. Robbins, Cracks and crazes: on calculating the macroscopic fracture energy of glassy polymers from molecular simulations. Phys. Rev. Lett. **89**(14), 148304 (2002)
53. S. Shenogin, R. Ozisik, Simulation of plastic deformation in glassy polymers: atomistic and mesoscale approaches. J. Polym. Sci. Part B Polym. Phys. **43**(8), 994–1004 (2005)
54. H.W. Spiess, Multidimensional solid state NMR: a unique tool for the characterisation of complex materials. Berichte der Bunsengesellschaft für physikalische Chemie, **101**(2), 153–168 (1997)
55. V. Srivastava, S.A. Chester, L. Anand, Thermally actuated shape-memory polymers: experiments, theory, and numerical simulations. J. Mech. Phys. Solids **58**(8), 1100–1124 (2010)
56. I. Strelnikov, N. Balabaev, M. Mazo, E. Oleinik, Analysis of local rearrangements in chains during simulation of the plastic deformation of glassy polymethylene. Polym. Sci. Ser. A **56**(2), 219–227 (2014)
57. G.R. Strobl, G.R. Strobl, *The Physics of Polymers*, vol. 2 (Springer, New York, 1997)
58. V. Sudarkodi, S. Basu et al., Investigations into the origins of plastic flow and strain hardening in amorphous glassy polymers. Int. J. Plast. **56**, 139–155 (2014)
59. L.R.G. Treloar, *The Physics of Rubber Elasticity* (Oxford University Press, Oxford, 1975)
60. M. Utz, A.S. Atallah, P. Robyr, A.H. Widmann, R.R. Ernst, U.W. Suter, Solid-state nmr investigation of the structural consequences of plastic deformation in polycarbonate. 1. Global orientational order. Macromolecules **32**(19), 6191–6205 (1999)
61. M. Utz, P. Robyr, U.W. Suter, Solid-state NMR investigation of the structural consequences of plastic deformation in polycarbonate. 2. Local orientational order. Macromolecules **33**(18), 6808–6814 (2000)
62. J.V. Van Dommelen, D. Parks, M. Boyce, W. Brekelmans, F. Baaijens, Micromechanical modeling of the elasto-viscoplastic behavior of semi-crystalline polymers. J. Mech. Phys. Solids **51**(3), 519–541 (2003)
63. I.M. Ward, J. Sweeney, *An Introduction to the Mechanical Properties of Solid Polymers* (Wiley, Hoboken, 2004)
64. I.M. Ward, J. Sweeney, *Mechanical Properties of Solid Polymers*. (Wiley, Hoboken, 2012)
65. M. Warren, J. Rottler, Deformation-induced accelerated dynamics in polymer glasses. J. Chem. Phys. **133**(16), 164513 (2010)
66. M. Warren, J. Rottler, Microscopic view of accelerated dynamics in deformed polymer glasses. Phys. Rev. Lett. **104**(20), 205501 (2010)
67. I.-C. Yeh, J.W. Andzelm, G.C. Rutledge, Mechanical and structural characterization of semicrystalline polyethylene under tensile deformation by molecular dynamics simulations. Macromolecules **48**(12), 4228–4239 (2015)
68. Q.-Y. Zhou, A. Argon, R. Cohen, Enhanced case-II diffusion of diluents into glassy polymers undergoing plastic flow. Polymer **42**(2), 613–621 (2001)

Chapter 4
Deformation-Induced Polymer Mobility and Sub-T$_g$ Bonding

Abstract This chapter discusses a new application of deformation-induced polymer mobility: sub-T$_g$, solid-state, plasticity-induced bonding. That is, active plastic deformation of polymer films held in contact leads to an enhanced mobility of macromolecules and interpenetration across the interface to cause bonding in time of the order of a fraction of 1 s. A short review of polymer adhesion due to conventional interdiffusion of macromolecules is carried out first, and then new results on plasticity-induced bonding are discussed. Contrasts between the conventional thermal welding of polymers and the newly reported deformation-induced bonding are discussed.

4.1 Conventional Polymer Adhesion Through Interdiffusion of Macromolecules

Polymer adhesion is encountered in several commercial applications and has been an active area of research. Voyutskii was the first to propose that diffusion was the major driving force for polymer adhesion [48]. Based on a plethora of research activity on polymer adhesion in the 1980s, most of the diffusion processes were assumed to be well understood [31]. The conventional wisdom was that if two pieces of a glassy polymer are brought into contact at a temperature above the T$_g$, along with the application of moderate contact pressures, polymer chains from the two sides interdiffuse on experimental timescales. The bonding results were explained on the basis that at higher temperatures (i.e., above the T$_g$) the polymeric chains acquire greater mobility, and on a macroscopic level there is a transition from a hard and relatively brittle state into a molten or rubber-like state. Thus, once the contact is established (and the material is heated above or close to its T$_g$) the interface broadens rapidly and interdiffusion begins. As a result of this macromolecular interpenetration, there is optical disappearance of cracks during healing [52] and development of strong bonds between the two surfaces after welding [9, 13, 14, 30, 35, 51, 54, 55]. The strength developing at an interface depends on the chemical structure of the polymers involved, their molecular weights and polydispersity, the geometry of the joint, and the method of testing. The

N. Padhye, *Molecular Mobility in Deforming Polymer Glasses*,
SpringerBriefs in Materials, https://doi.org/10.1007/978-3-030-82559-1_4

strength of adhesion has been reported to be a function of temperature, time of healing, and contact pressure, and the healing process continues until the interface acquires the bulk properties. Typically, for times shorter than the bulk reptation time, the interface toughness (G_c) and shear strength (σ_{max}) show a monotonic time-dependent growth as $G_c \sim t^{1/2}$ and $\sigma_s \sim t^{1/4}$ [18, 34, 53, 55], along with dependence on the molecular characteristics of the polymers involved. Symmetric amorphous interfaces (i.e. same polymers on both sides) provide an equivalent molecular flux in opposite directions and tend to develop stronger bonds with that of immiscible polymers. The need for contact pressure during healing applications has been reported to be necessary to promote intimate contact between the interfaces brought together into proximity so that interdiffusion can occur.

In a series of recent computer simulations using coarse-grained models with bond-breaking potentials, Ge et al. [23–25] examined the bonding between polymer layers as a function of annealing time above the T_g. Their work revealed that the key to bonding between polymer films is the reestablishment of the entanglement network across the interface. Their findings were consistent with the existing experimental results in terms of scaling laws for the growth of maximum shear strength of the interface and fracture toughness as a function of healing time. The increase in entanglement density across the interface was shown to grow at the same rate with which the interfacial width broadens, also observed experimentally, and that this rate was largely determined by reptation dynamics, albeit with some deviations attributed to relaxation of the perturbed conformations of chains typically found at the surface of a film. All such welding simulations were undertaken at an increased temperature without any imposed deformation.

Reptation dynamics of polymer chains [17], as discussed in Chap. 1, have been used to describe the scaling laws for interfacial strength during interdiffusion bonding. For studies on establishing the reptation-like behavior during interdiffusion see [34, 36, 50]. We will discuss the basic ideas behind the role of reptation during interdiffusion, and later utilize them to draw contrast with deformation-induced bonding. The growth of the number of bridges $p(t)$ (pieces of polymer chain per unit area spanning interface junction) during interdiffusion, as a function of time, was analyzed by de Gennes [17] based on simple arguments, and in more detail by Prager and Tirrell [45]. A time-dependence of $p(t) \sim t^{1/2}$ was suggested. As an illustration, a bridge across the interface is shown in Fig. 4.1. Although there is no

Fig. 4.1 Polymer-chain entanglement and formation of bridges across the interface

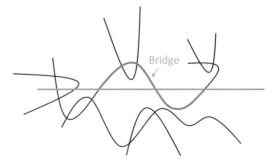

Fig. 4.2 Reptation of a
polymer chain

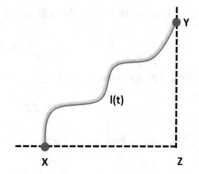

straightforward relationship between the number of bridges and fracture energy, a
simple model based on plastic deformation at the onset of fracture does lead to G_c
$\sim p(t)$, discussed in [18]. This also explained the earliest experiments done in [30],
in which a dependence of G_{1c} on $t^{1/2}$ was reported.

For the sake of illustration, we can employ scaling arguments and establish how
the interface growth $x(t) \sim t^{1/4}$ can be derived using reptation dynamics. According
to Fig. 4.2, we can consider the diameter of the tube, in which a particular chain is
constrained, as d. Let the length of the contour be $l(t)$ during reptation. Then,

$$l(t) \approx (D_t t)^{1/2},$$

where D_t is the diffusion coefficient in the tube. The chain is assumed to be an ideal
Gaussian, so we can discuss XY as an end-to-end vector with N segments each of
size d (the representative tube diameter and diameter of each monomer coil is d).
Thus,

$$\|\mathbf{XY}\|^2 \approx Nd^2, \tag{4.1}$$

where $\| \cdot \|$ indicates the scalar norm of the vector argument. We also know that

$$l(t) \approx Nd. \tag{4.2}$$

Using Eqs. 4.1 and 4.2

$$\|\mathbf{XY}\|^2 \approx \frac{l(t)}{d}d^2 \approx l(t)d \tag{4.3}$$

or

$$\|\mathbf{XY}\| \approx (l(t)d)^{1/2} \approx ((D_t t)^{1/2}d)^{1/2} \sim t^{1/4}. \tag{4.4}$$

If we argue that chain diffusion is isotropic, then we can say that

$$\|\mathbf{XZ}\|^2 \approx \|\mathbf{XY}\|^2, \tag{4.5}$$

and we know that

$$\|\mathbf{XZ}\|^2 + \|\mathbf{ZY}\|^2 = \|\mathbf{XY}\|^2. \tag{4.6}$$

Thus, using Eqs. 4.4, 4.5 and 4.6, we get

$$\|\mathbf{XZ}\| \approx \|\mathbf{XY}\| \sim t^{1/4}. \tag{4.7}$$

Thus, the interface width grows with a time-dependent scaling of $t^{1/4}$ due to reptation. Typically, the diffusion process is assumed to be complete, i.e., the interface is fully healed when the chains have crossed a distance of the order of the radius of gyration. This is also the time when healing is complete. For times greater than the reptation time, all the correlated effects are lost, and the progress of the front can be described according to normal Fickian diffusion. That is, at times greater than the reptation time, the displacement of monomers and the center of mass of the chain show a time dependence of $t^{1/2}$. Now, we can discuss how the diffusion welding is argued to occur through reptation processes, and hairpin-like processes would have much less likelihood (Fig. 4.3). We know that the probability distribution for a Gaussian coil is given as

$$P(R, N) = \frac{1}{\sqrt{2\pi N}} e^{-R^2/2N}.$$

Here, we have assumed $a = 1$. If we consider a total chain length of N_e, and are interested in a "hairpin" with n segments, then

$$R^2 = n, \tag{4.8}$$

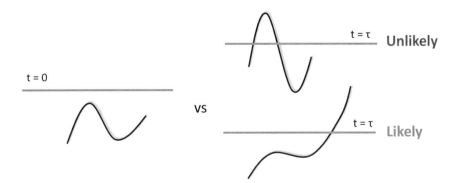

Fig. 4.3 Hairpin vs. reptation process during healing of the interface

and the probability for this "hairpin" is given as

$$P = \frac{1}{\sqrt{2\pi N_e}} e^{-R^2/2N} = \frac{1}{\sqrt{2\pi N_e}} e^{-n/2N_e}. \tag{4.9}$$

The probability that a "hairpin" process will be sampled decreases as the contour length "n" or distance it sweeps (R^2) increases. This is why "hairpin" processes are disfavored [19], and why the motion of a polymer chain, constrained along the tube contours is the primary mechanism of mobility across the interface. This analysis is valid for the behavior of an ideal polymer chain, however, it is a useful approximation to derive some physical understanding.

Some scholars have investigated the bonding of polymers below the T$_g$, such as see [4, 6, 46]. In each case, application of moderate contact pressure has been reported to ensure intimate contact between surfaces. In [3, 5], researchers reported the adhesion of polymers below the T$_g$ over relatively long intervals (minutes-to-hours) under moderate contact pressures. Those studies claimed that the surface layer of the polymers are viscoelastic (or in a rubbery state), the T$_g$ of the surface layers is lower than the bulk T$_g$ and, hence, there is naturally enhanced molecular mobility at the polymer free-surface due to which interdiffusion occurs (as discussed in Chap. 2). However, the role of mechanically activated molecular mobility aiding the adhesion due to interdiffusion above or below T$_g$ has not been discussed.

Next, we will discuss some recent work [42, 44] on how active plastic deformation of glassy polymeric films, held in intimate contact, triggers the requisite molecular-level rearrangement in facilitating interpenetration of polymer chains across the interface, which leads to bonding of the order of 1 s.

4.2 Deformation-Induced Sub-T$_g$, Solid-State Bonding

Figure 4.4 contrasts the underlying idea of polymer self-adhesion through inter-diffusion and newly proposed deformation-induced bonding. Here, amorphous polymeric films were prepared from solvent casting and bonded in the solid state at ambient temperatures, approximately 60 K, below their T$_g$, in time on the order of a *second*, by subjecting them to active compressive plastic deformation. Despite the glassy-regime, bulk plastic deformation triggered the requisite molecular mobility of polymer chains to cause chain interpenetration across the interfaces. Quantitative levels of adhesion and morphology of the fractured interfaces validated the proposed model of sub-T$_g$, plasticity-induced, molecular mobilization causing bonding. No-bonding outcomes (i) during compression of films in a near hydrostatic setting (which does not cause plastic flow), and (ii) between an "elastic" film and a "plastic" film, further established the explicit role of plastic deformation in this process. Procedures for preparing film samples, film characterization, bonding results, and general discussions are presented next.

'Interdiffusion' above T$_g$

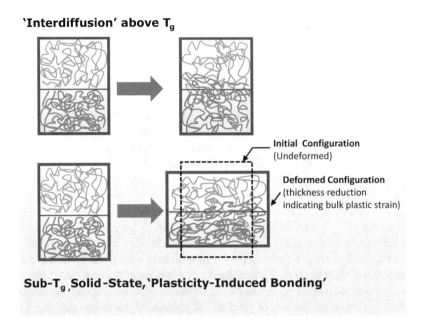

Initial Configuration
(Undeformed)

Deformed Configuration
(thickness reduction
indicating bulk plastic strain)

Sub-T$_g$ Solid-State, 'Plasticity-Induced Bonding'

Fig. 4.4 Comparison of polymer self-adhesion through diffusion at temperatures near or above T$_g$ and, newly proposed sub-T$_g$, solid-state, plasticity-induced bonding, where bulk plastic deformation triggers the requisite molecular mobility for chain interpenetration across the interfaces

4.2.1 Preparation of Polymeric Films

Polymeric films were prepared from solvent casting (Fig. 4.5) using the base polymer hydroxypropyl methylcellulose (HPMC) and a compatible plasticizer polyethylene glycol (PEG-400). Their molecular structures are shown in Fig. 4.6. HPMC with trade name METHOCEL, in the grades E3 and E15, was obtained from Dow. PEG-400 was purchased from Sigma-Aldrich. Appropriate amounts of E3, E15 and PEG were mixed in desired amounts with ethanol and water, and a homogeneous solution was obtained through mixing with an electric stirrer for over 24 h. After completion of the blending process, the solution was carefully stored in glass bottles at rest for 12 h to remove air bubbles. Solvent casting was carried out using a casting knife applicator from Elcometer on a heat-resistant Borosilicate Glass. All steps were carried out in a chemical laboratory with ambient conditions of $20 \pm 2\,°C$ and R.H. $20 \pm 5\%$.

HPMC is an uncrosslinked polymer and shows excellent film formability due to its underlying cellulose structure, in which all functional groups lie in equatorial positions, causing the molecular chain of cellulose to extend in a more-or-less straight line and easing film formation. Table 4.1 shows the sample weights of the contents used in preparation of the solutions. Films were made through casting and

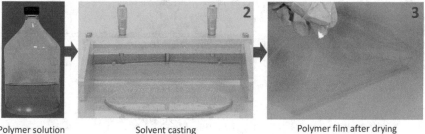

Polymer solution   Solvent casting   Polymer film after drying

Fig. 4.5 Steps involved in preparation of polymer films through solvent casting (1) homogeneous solution of polymer and plasticizer in ethanol and water, (2) spreading of solution on glass surface through a knife, (3) evaporation of solvents and formation of glassy film after drying

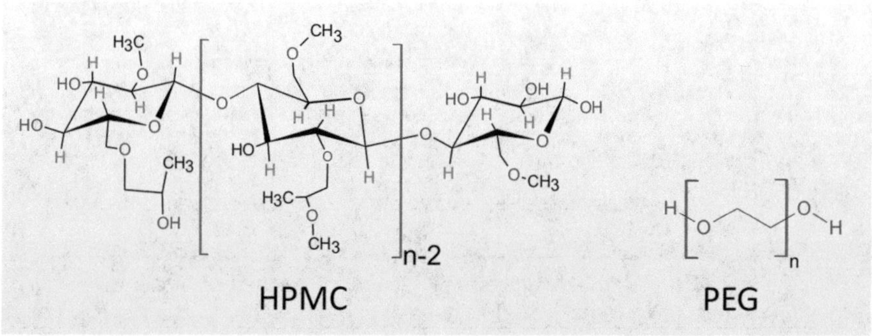

Fig. 4.6 Molecular structures of hydroxypropyl methylcellulose (HPMC) and polyethylene glycol (PEG)

Table 4.1 Formulations employed in making polymer films from HPMC E3 and E15 with different levels of plasticizer (amounts have been rounded off to the nearest gram)

| | Composition | | | | |
Polymer film	E3 (g)	E15 (g)	Water (g)	EtOH (g)	PEG (g)
E3/E15 in 1:1-0% PEG	15	15	96	96	0
E3/E15 in 1:1-28.5% PEG	15	15	96	96	12
E3/E15 in 1:1-42.3% PEG	15	15	96	96	22
E3/E15 in 1:1-59.5% PEG	15	15	96	96	44
E3-alone-42.3% PEG	30	0	96	96	22
E15-alone-42.3% PEG	0	30	96	96	22

drying of the prepared solutions. As seen in Table 4.1, films are referred to based on the amounts of E3, E15 and Wt.% of PEG-400. For example, E3/E15 in 1:1-42.3% PEG implies that E3 and E15 are present in a one-to-one ratio and the Wt.% of PEG in the film is 42.3% (because 22 g of PEG in 15 g of E3 and 15 g of E15 corresponds to 42.3% of the total mass).

Karl Fischer titration was carried out to determine the residual moisture content in the films after drying, because water can act as a plasticizer and effect the physical

Table 4.2 Measurement of residual moisture in films after drying through Karl Fischer titration. %Wt. indicates residual moisture in the films after drying

Polymer film	Residual H$_2$O (%Wt.)
E3/E15 in 1:1-0% PEG	3.70
E3/E15 in 1:1-28.5% PEG	7.21
E3/E15 in 1:1-42.3% PEG	4.29
E3/E15 in 1:1-59.5% PEG	2.45
E3-alone-42.3% PEG	2.92
E15-alone-42.3% PEG	4.54

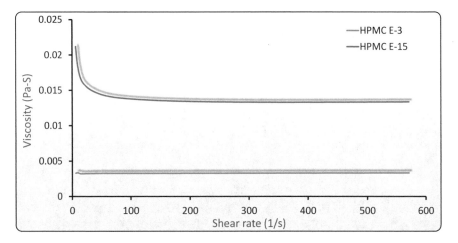

Fig. 4.7 Viscosity curves of 2% aqueous solution of METHOCEL-E3 and METHOCEL-E15

and mechanical properties. Dimethyl sulfoxide (DMSO) was used as a reagent for dissolving films. DMSO (ACS reagent grade) was purchased from Sigma Aldrich. The solution of the dissolved film in DMSO was fed into a Karl Fischer titrator, and the residual moisture was estimated. Table 4.2 shows the average amounts (repeated thrice) of the estimated residual moisture contents in the films.

4.2.2 Properties and Characterization of Polymeric Films

To estimate the molecular weight of E3 and E15, a standard procedure prescribed by Dow [40] was followed. Viscosity measurements for a 2% aqueous solution of E3 and E15 were carried out using an HR-3 Hybrid rheometer. The viscosity curves for E3 and E15 are shown in Fig. 4.7. The E3 solution showed negligible rate-dependence, with the viscosity lying in the range 3–4 mPa-s. E15 solution showed some rate dependence, with viscosity lying in the range 14–21 mPa-s. The Dow manual [39], stated that the viscosity for a 2% aqueous solution of grades E3, E5, E6, E15, and E50 as 3, 5, 6, 15, and 50 mPa-s, respectively. In another Dow manual [26], the ranges for the viscosity of a 2% Wt. solution of E3 and E15 were specified as 2.4–3.6 mPa-s and 12–18 mPa-s, respectively. The obtained measurements overlapped well within these specifications. If we chose a representative viscosity of 3.8 mPa-s for E3 and 16 mPa-s for E15 then, based on the viscosity and molecular weight relationship from [39], the number average molecular weight (M_n) for E3 and E15 would be 8200 and 20, 000, respectively.

A detailed molecular characterization of METHOCEL cellulose ethers, presented in [32], also led to estimation of weight-average molecular weight (M_w) and number average molecular weight (M_n) as: (i) E3: $M_n = 8100$ and $M_w = 20, 300$, with $M_w/M_n = 2.5$, and (ii) E15: $M_n = 24, 800$ and $M_w = 60, 300$, with $M_w/M_n = 2.4$. Such estimations are consistent with those found here. In the same study, the degree of polymerization (DP) was reported as: (i) E3: DP= 77, and (ii) E15: DP= 296, and the weight-average radius of gyration (R_{gw}) as: (i) E3: $R_{gw} = 7.4$ nm, and (ii) E15: $R_{gw} = 15.1$ nm.

Based on these estimates, it became clear that E3 and E15 were low or moderate-molecular weight grades of HPMC, respectively. This is a critical observation because to achieve deformation-induced bonding, the chains must move across the interface and form entanglements upon plastic deformation. Presence of the chain-ends on the free surface is likely to enable such bonding, and polymers of low-to moderate-molecular weight will have relatively more chain ends compared with those of very high-molecular weight grades.

PEG-400 acts as a compatible plasticizer for HPMC. Inclusion of a plasticizer within the polymer matrix enhances its ductility. Plasticizers work by embedding themselves between the chains of polymers and spacing them apart by increasing the free volume. By dissolving and mixing intimately, PEG molecules disrupt the secondary bonds between polymer chains. Figure 4.8 illustrates this effect schematically.

Figure 4.9 shows the tensile true stress-strain curves of films made from E3/E15 in 1:1 ratio with 0%, 28.5%, 42.3% and 59.5% PEG concentration. The plasticization effect of increasing PEG (Wt.%) was evidenced by lowering of the yield stress and increase in failure strain. The maximum failure strain was noted for 42.3% PEG film. Clearly, all films containing PEG demonstrated large plastic flow that was absent in 0% PEG film. Films with 42.3% PEG showed maximum

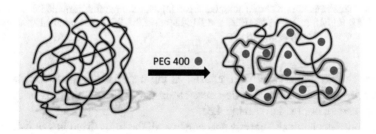

Fig. 4.8 Role of a plasticizer (PEG-400)

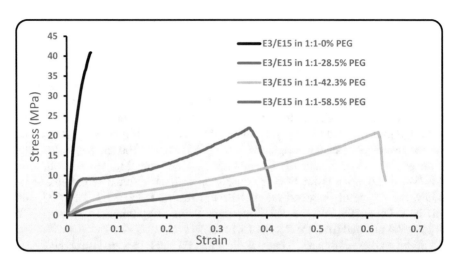

Fig. 4.9 Effect of PEG-400 on tensile true stress-strain behavior of polymeric films. The tensile tests were carried out at ambient temperature and a nominal strain rate of 0.0025 s^{-1}

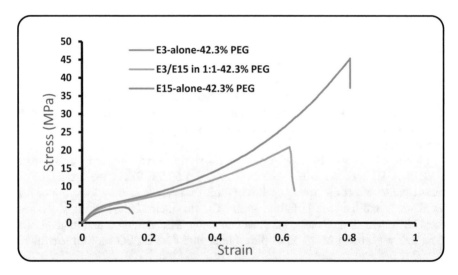

Fig. 4.10 True stress strain curves at ambient conditions for three film formulations: E3-alone-42.3% PEG, E3/E15 in 1:1-42.3% PEG and E15-alone-42.3% PEG. The nominal strain rate for tensile testing was chosen as 0.0025 s^{-1}

ductility characteristics, and were chosen as the material candidate for plasticity-induced bonding. The true tensile stress-strain curves for three films with 42.3% PEG are shown in Fig. 4.10 (Table 4.3).

Dynamic Mechanical Analysis was done on all the films listed in Table 4.1 using a TA Q800 instrument. A temperature sweep was undertaken at 1 Hz frequency. Figures 4.11, 4.12, 4.13, 4.14, 4.15, and 4.16 show the plots of loss modulus, storage

Table 4.3 Modulus and yield strength of films with 42.3% PEG, based on the nominal strain rate of 0.0025 s^{-1} in tensile testing (corresponding to Fig. 4.10)

	Modulus (MPa)	Yield strength (MPa)
E3-alone-42.3% PEG	82.3	4.12
E3/E15 in 1:1-42.3% PEG	74.7	3.72
E15-alone-42.3% PEG	59.1	2.96

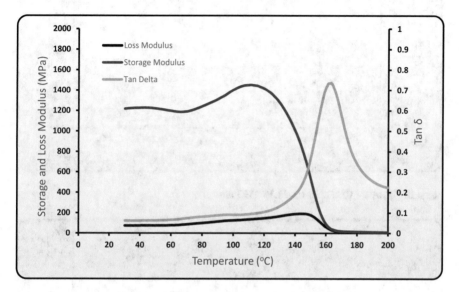

Fig. 4.11 DMA for E3/E15 in 1:1-0% PEG film

modulus and tan δ for six different films. The T$_g$ was determined by the peak in tan δ. For E3-alone-42.3% PEG, the T$_g$ was estimated to be 72 °C, and for E15-alone-42.3% PEG and E3/E15 in 1:1-42.3% PEG the T$_g$ was estimated as 78 °C. Inclusion of PEG evidently lowered the T$_g$ and broadened the temperature range over which the glass transition occurred. Broad tan δ curves were interesting and suggested that the distinction between the glassy state and rubbery state was not as sharp as noted for conventional engineering thermoplastics. This suggests the plausibility that, even in solid state, well below T$_g$ (which is marked by the peak in the tan δ curve), it is plausible to strongly activate polymer mobility, and, as such, this is a special type of glass. The physical state is arguably solid because films exhibited a clear yield strength (and hardness, which is discussed later). However, the molecular level relaxation mechanisms that would be induced in this type of glass during plastic deformation are not known.

HPMC is a cellulose derivative and exists in an amorphous form. To verify its amorphous characteristic, an X-ray diffraction study was carried out on E3-alone-42.3% PEG, E15-alone-42.3% PEG and E3/E15 in 1:1-42.3% PEG films, as shown in Figs. 4.17, 4.18 and 4.19, respectively.

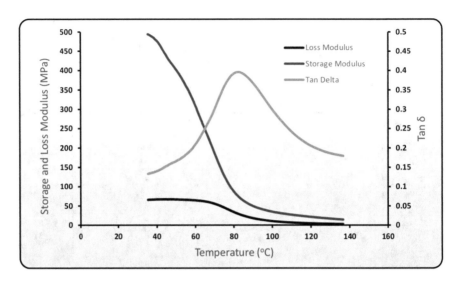

Fig. 4.12 DMA for E3/E15 in 1:1-28.5% PEG film

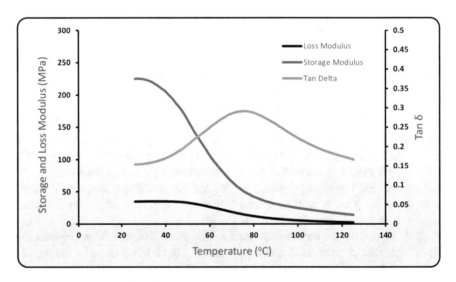

Fig. 4.13 DMA for E3/E15 in 1:1-42.3% PEG film

As expected, a diffused pattern without any sharp peaks was obtained in all films, indicating an absence of crystallinity. The X-ray diffraction patterns were recorded on a PANalytical X'Pert PRO Theta/Theta Powder X-Ray Diffraction system with a Cu tube and X'Celerator high-speed detector.

Nanoindentation experiments were undertaken on E3/E15 in 1:1-0% PEG, E3/E15 in 1:1-42.3% PEG, E3-alone-42.3% PEG and E15-alone-42.3% PEG films.

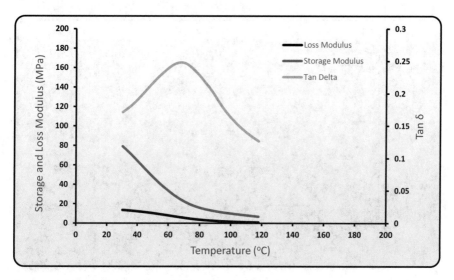

Fig. 4.14 DMA for E3/E15 in 1:1-58.5% PEG film

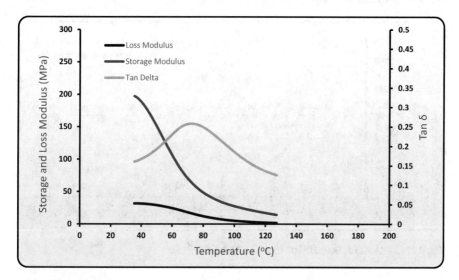

Fig. 4.15 DMA for E3-alone-42.3% PEG film

The experiments were carried out in force-controlled mode with a maximum force of 2000 μN and 300 μN for 0% PEG and 42.3% PEG films, respectively. Triboindenter Hysitron equipment was used for experimentation. A larger load for the 0% PEG film was chosen to activate sufficient plastic indentation so that its hardness could be measured. A Berkovich indenter with a root radius of 150 nm was used. The load versus displacement curves for all the films are shown in Fig. 4.20. The film with 0% PEG showed a relatively large indentation force and large elastic

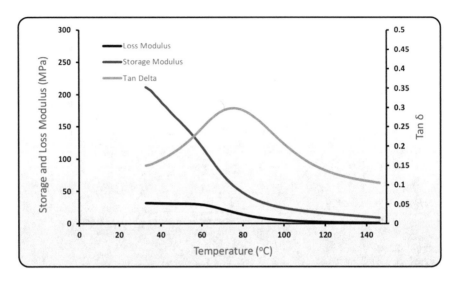

Fig. 4.16 DMA for E15-alone-42.3% PEG film

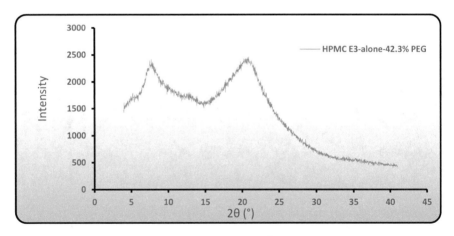

Fig. 4.17 XRD results from HPMC E3-alone-42.3% PEG film

recovery, whereas films with 42.3% PEG films showed little elastic recovery and large residual indentation depth. Based on these behaviors, on a relative basis, we called the 0% PEG film an "elastic" film and the 42.3% PEG film a "plastic" film.

Using the Oliver-Pharr method [41], we estimated the hardness from the nanoindentation tests. The inferred values for hardness (in MPa) for E3/E15 in 1:1-0% PEG, E3/E15 in 1:1-42.3% PEG, E3-alone-42.3% PEG, and E15-alone-42.3% PEG films were 144.0, 10.83, 10.151, and 11.48, respectively.

AFM images were obtained with a Dimension 3100 XY Closed-Loop Scanner (Nanoscope IV, VEECO) equipped with NanoMan software. Height and phase

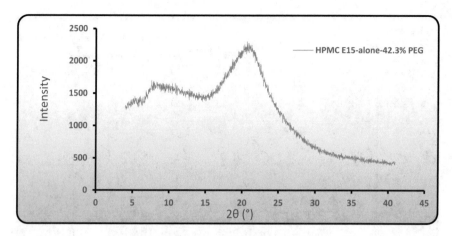

Fig. 4.18 XRD results from HPMC E15-alone-42.3% PEG film

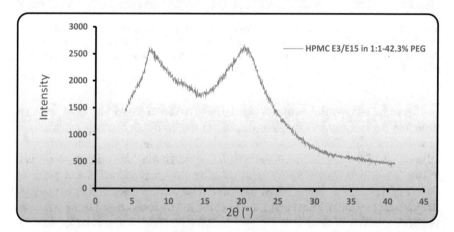

Fig. 4.19 XRD results from HPMC E3/E15 in 1:1-42.3% PEG film

images were obtained in tapping mode in ambient air while using silicon tips (VEECO). Figure 4.21 shows sample AFM scans of 5 μm × 5 μm area on the top surface of three films having 42.3% PEG, along with the average roughness given by $R_a = \frac{\sum_1^n |y_i|}{n}$. The top surfaces of films exhibited nano-scale roughness of order 6.91–22.7 nm.

4.2.3 Deformation-Induced Bonding Experiments

All bonding experiments were carried out at ambient conditions using a roll-bonding machine that was custom designed, or in a lap-shear specimen geometry.

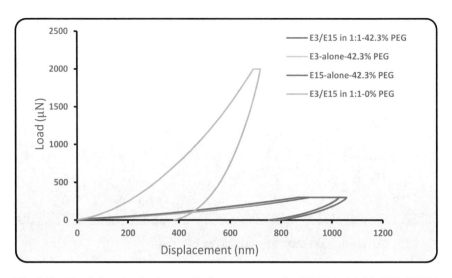

Fig. 4.20 Nanoindentation load versus displacement curves for E3/E15 in 1:1-0% PEG, E3/E15 in 1:1-42.3% PEG, E3-alone-42.3% PEG and E15-alone-42.3% PEG films. Indentation experiments were carried out in load controlled mode with chosen peak loads up to 300 μN and 2000 μN for films with 42.3% PEG and 0% PEG, respectively

Figure 4.22 shows the computer aided design model of the roll-bonding machine designed for this work. Complete details on the cold-roll bonding machine are provided in [42]. The machine can achieve different levels of plastic strain by adjustment of the gap between the rollers and monitoring the compression load. Plastic strain is measured in terms of effective thickness reduction and given as $|t_1 - t_2|/t_1$, where t_1 is the initial total thickness of film stack and t_2 is the total thickness at the outlet. The angular speed of the rollers is controlled using a stepper motor. The radius of the rollers (R_{roller}) is 100 mm, which is much larger than the total initial thickness of a film-stack (t_1, typically less than 1 mm). The incoming stack of film behaves like a thin strip, and through-thickness plastic deformation can be triggered under such conditions. From the kinematics of rigid-plastic rolling of a "thin strip" [29], the time spent during active plastic deformation can be estimated as:

$$\tau = \frac{\sqrt{R_{roller}(t_1 - t_2)}}{V_2}. \tag{4.10}$$

In Eq. 4.10, t_1 is the initial thickness of the film stack, t_2 is the thickness of film-stack at the exit, and V_2 is the linear speed at the exit. For $V_2 = 5.23$ mm/s, t_1=0.6 mm and t_2=0.45 mm (indicating 25% plastic strain), the time spent by a material element under the roller would be approximately 0.74 s. For different levels of plastic strains, estimates of times spent under the rollers are given in Table 4.4.

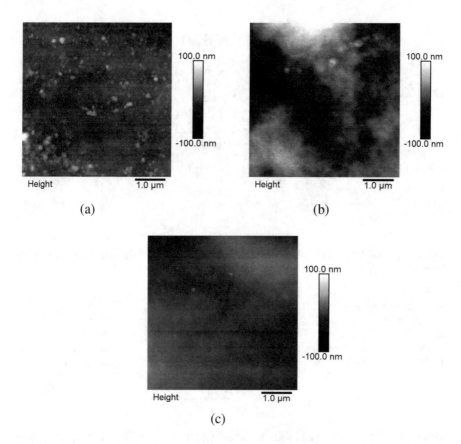

Fig. 4.21 Measurement of nano-roughness using atomic force microscopy. (a) Top surface of E3/E15 in 1:1-42.3% PEG film, $R_a = 6.91$ nm. (b) Top surface of E3-alone-42.3% PEG film, $R_a = 22.7$ nm. (c) Top surface of E15-alone-42.3% PEG film, $R_a = 8.63$ nm

Stacks of six film layers (thickness of each layer $\sim$100 μm) were fed through a designed roll-bonding machine to achieve different levels of plastic deformation, and peel tests were carried out to measure mode-I fracture toughness (G_c [J/m^2]). This is shown schematically in Fig. 4.23. G_c represents the work done per unit area for debonding the interface during a peel test. See [43] for complete details on peel-tests. For all experiments, the rollers were operated at an angular speed of 0.5 rpm, leading to an exit speed of 5.23 mm/s. The error bars in G_c are based on the variation in peeling force once steady state was achieved.

Lap specimens were prepared, using upsetting, to measure shear-strength (σ_s [MPa]), shown in Fig. 4.24. Here σ_s indicates the maximum shear stress sustained by the bonded interface before failure when loading the bonded strips in tension. Effective thickness reduction during bonding was taken as a measure of plastic strain. Preparation of lap specimens and shear-strength measurements were carried out using an Instron testing machine. A lap joint was assembled between two

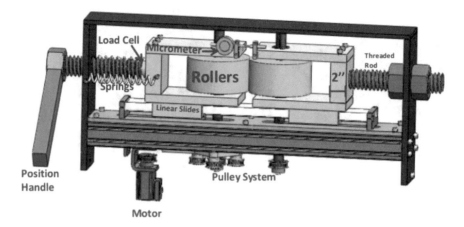

Fig. 4.22   A roll-bonding machine employed to carry out sub-T$_g$, solid-state, plasticity-induced bonding

Table 4.4 For a given exit speed ($V_2 = 5.23$ mm/s) and an initial thickness $t_1 = 0.6$ mm, estimates of time spent under the roller bite during plastic strain are given

Plastic strain	Final thickness (t_2) (mm)	Time (τ) (s)
0.05	0.57	0.33
0.1	0.54	0.47
0.15	0.51	0.57
0.20	0.48	0.66
0.25	0.45	0.74

layers, each layer having an initial thickness about 100 μm. The overlapping region comprised an area of approximately 5×5 mm^2. A cross-head speed of 0.5 mm/min was chosen to apply the desired compression load on the overlapping area. The sample was bonded plastically by pressing between two parallel (accuracy ~ 1 μm) flats. The resulting lap joint was tested for shear-strength in tension mode at a cross-head speed of 15 mm/min.

For both the roll-bonded and lap shear-strength specimens, the measurements for the adhesion were carried on the interface formed between bonded top-top surfaces of films.

4.3   Results and Discussions on Deformation-Induced Bonding

Figure 4.25 shows a snapshot of several layers of the film (E3/E15 in 1:1-42.3% PEG) with an initial-thickness of 0.024″ (0.609 mm) undergoing roll-bonding through active plastic deformation, with a final thickness reduced to 0.021″ (0.533 mm).

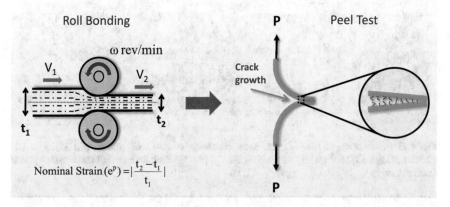

Fig. 4.23 Roll-bonding is achieved by passing a stack of film layers with a total initial thickness t_1 under compression rollers to yield a final-thickness t_2. A peel test is carried out on roll-bonded stock at the middle interface

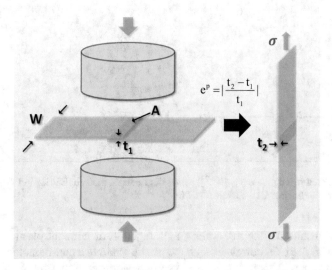

Fig. 4.24 Lap specimens were prepared between two film layers by applying compression loads on an overlapping area (A). Lap shear-strength measurements were done in a tensile mode

Stack of film layers   Multiple layers undergoing sub-T_g, solid-state, roll-bonding   Roll-bonded layers
initial thickness 0.024'' (0.60 mm)   (final thickness 0.021'' (0.53 mm)

Fig. 4.25 Illustration of sub-T_g, solid-state, plasticity-induced roll-bonding of E3/E15 in 1:1-42.3% PEG films approximately 60 K below the T_g. For this case, the nominal compressive plastic thickness strain was $e_p = (t_1 - t_2)/t_1 = 11.7\%$

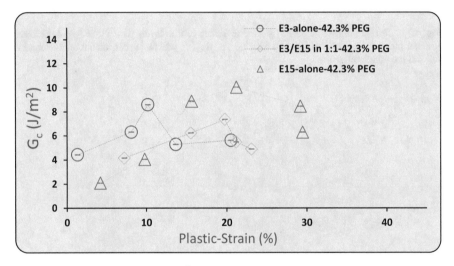

Fig. 4.26 Fracture toughness (G_c [J/m^2]) versus plastic strain plots for E3/E15 in 1:1-42.3%PEG, E3-alone-42.3%PEG and E15-alone-42.3%PEG

The G_c results for the three films, as a function of nominal plastic strain, are shown in Fig. 4.26. G_c correlates with the plastic strain in a non-monotonic fashion, first increasing and then decreasing. The adhesion between two interfaces held together by van der Waals forces, hydrogen bonds, or chemical bonds can only give G_c values in the range of 0.05 J/m^2, 0.1 J/m^2 and 1.0 J/m^2, respectively [20].[1] The surface energy of glassy polymers itself is quite small [8] (on the order 0.08

---

[1]The largest bond energies associated are of the order 1000 kJ/mol (for a carbon atom double-bonded to an oxygen atom, or a carbon atom triple-bonded with another carbon atom), which equates to approximately 10 ev, and, if we consider G_c due to 10 ev per $(3 \text{ Å})^2$, it amounts to approximately 10 Joules/m^2. In the present system breaking of such bonds is not easy, and huge forces associated during this kind of bond scission will first lead to chain drag out, with the associated plastic work contributing to measured G_c values.

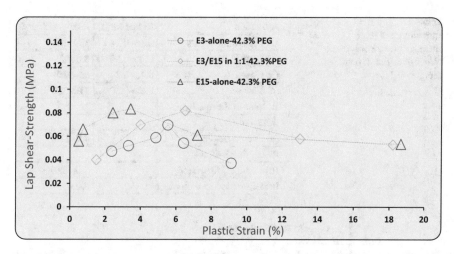

Fig. 4.27 Lap shear-strength (σ_s [MPa]) versus plastic strain plots for E3/E15 in 1:1-42.3%PEG, E3-alone-42.3%PEG and E15-alone-42.3%PEG

J/m^2), and therefore negligible adhesion is noted when two such surfaces are simply brought into molecular proximity. However, glassy polymers can exhibit larger fracture toughness due to irreversible deformation of the macromolecules [10]. The quantitative levels of G_c obtained here can only be attributed to the irreversible processes of chain pull-outs, disentanglement and/or scissions, during debonding; and this could only happen if plasticity-induced molecular mobilization and chain-interpenetration led to bonding. Even polymer adhesion leading to G_c values as low as 1.2 J/m^2 [4], and 2.0 J/m^2 [49] have been attributed due to irreversible mechanisms of chain pull-outs during fracture. Other mechanisms of adhesion such as acid-base interaction, capillary effect, electrostatic forces and/or any other conceivable mechanisms do not apply in this context (for a detailed discussion on types of forces giving rise to adhesion, see [33, 38]). Van der Waals bonds nor even the strongest ionic or covalent bonds alone can give a very significant fracture toughness unless accompanied by irreversible work during interface failure. The lap shear-strength of the specimens shown in Fig. 4.27, also support plasticity-induced bonding.

The types of intermolecular and interatomic attractions are shown in Table 4.5. For this discussion, we are interested in the forces acting at a molecular level. Figure 4.28 shows the universal nature of these attractions. Thus, long-range attraction forces can be of two types: (a) Coulombic or electrostatic forces, or (b) van der Waals forces. Forces of type (a) arise due to Coulombic interactions between charges, permanent dipoles and quadrapoles, whereas those of type (b) include

Table 4.5 Types of attractions and their characteristics. For typical values of different bond energies see [7]

Attraction	Range (nm)	Energy (kJ/mol)
I. Inter-atomic bonds	0.15–0.24	335–1050
(a) Ionic H-bond	0.26–0.30	8.0–42
(b) Covalent	0.15–0.24	63–920
(c) Metallic	0.26–0.30	110–350
II. Inter-molecular bonds		
(indefinite range)		
(a) Dipole-dipole		4.0–21.0
(excluding H-bonds)		
(b) London (dispersion)		4.0–4.2
(c) Dipole-induced dipole		≈2.0

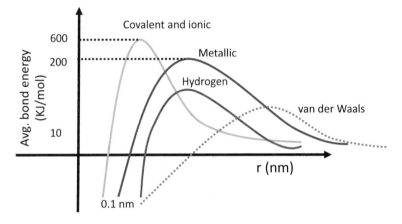

Fig. 4.28 Energies associated with different bonds as function of a distance

dispersion and polarization forces that arise from dipole moments induced in atoms and molecules by the electric fields of nearby changes and permanent dipoles.[2]

In the current case, PEG may show H-bonding character (with HPMC). If H-bonds act across two interfaces with one H-bond per (3Å)^2 area, we would expect toughness on the order of a fraction 1 J/m^2. Thus, solely H-bonds acting across an interface cannot lead to toughness up to 10 J/m^2, as reported here. For the polymeric films considered in this chapter, H-bonding cannot contribute to toughening effect unless a polymer chain embeds itself across the interface. The weak effect of only

---

[2]The Lennard-Jones (LJ) potential models the interaction between one pair of neutral atoms. The LJ potential is given as $u = (\frac{A}{r^{12}} - \frac{B}{r^6})$, where A and B are constants. The intermolecular pair potential and van der Waals interaction is frequently represented by the LJ potential. At larger distances the nature is attractive, and at shorter distances it is repulsive. This behavior also explains why any forces at the sub-atomic level cannot be relevant because, as molecules or atoms are brought extremely close, there will be a very strong repulsion, and so forces acting within the nucleus of an atom are not relevant.

Fig. 4.29   H-bonding across the interface due to PEG molecules, without crossing of any polymer chains across the interface

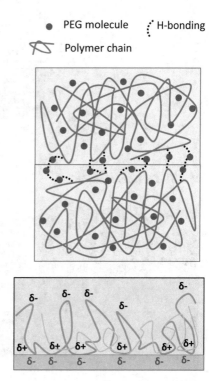

Fig. 4.30   Adhesion due to electrostatic interactions

H-bonds across the interface are shown schematically in Fig. 4.29. For the density of polymers with HPMC+PEG on both sides of the interface, a single PEG molecule on the surface can participate through one (or two) H-bond(s) with either a PEG on the other surface, or with some functional group of the HPMC molecule. Unless the HPMC chain is "injected" on the other side, the molecules in the sub-surface cannot show any H-bonding across the interface, and we know that whichever H-bonds can be formed with PEG on the surface will lead to toughness of the order of a fraction of 1 J/m^2. (This is under assumption that the PEG molecule is extremely small, which it is, compared with large HPMC molecules.) Using bond energies for van der Waals and ordinary chemical bonds, we arrive at similar conclusions.[3]

Figure 4.30 shows, schematically, the electrostatic forces acting between two interfaces. The mechanism of electrostatic forces of adhesion between symmetric junctions (or if there is no difference in redox potential) is inapplicable. Electrostatics forces can become applicable in unsymmetric junctions (i.e., with suitable electron donors and acceptors crafted on each side of the interface [20]). Lastly,

---

[3]In relation to our work, even if one speculates that during plastic deformation PEG forms an "interlayer", which could have an important role in adhesion then, based on the same arguments, as presented here, any such (remotely) possible layer is unlikely to lead to toughness up to 10 J/m^2. Lastly, new chemical bonds are formed during plastic deformation in the current process.

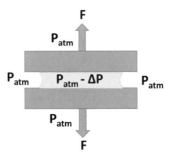

Fig. 4.31 Capillary adhesion

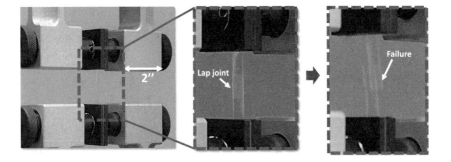

Fig. 4.32 A test specimen for lap shear strength in a tensile tester

the mechanism of capillary adhesion, shown in Fig. 4.31, is also not applicable to the reported deformation-induced bonding. Even if a PEG molecule forms an intermediate layer to show a capillary effect, it can only lead to high a "breaking" force at the onset of the failure, but cannot provide the reported levels of toughness, or explain the morphology of the fractured surfaces (as seen shortly).

The shear strengths (σ_s) plots, shown in Fig. 4.27, also exhibited a non-monotonic correlation with bonding plastic strain. Quantitative levels of σ_s values reported here compare with those in [5], where adhesion due to interdiffusion of chains below the bulk T$_g$, pressed together over long times, was reported. A snapshot of the shear-strength test is shown in Fig. 4.32. The peak force before failure (F_{max}), divided by the bonded area (A), was taken as the lap shear-strength (σ_s). Figure 4.33 shows the force versus displacement during measurement of lap shear strength for a particular material and bond condition.

The reported levels of bulk plastic strains also rule out mechanical interlocking of asperities as a potential cause of adhesion. At the levels of plastic strains reported here, the surface asperities will necessarily plastically flatten out. As stated previously, the pre-bonding surface characterization of films through AFM revealed nano-scale roughness (R_a) of the order of 10–26 nm. Conversely, the levels of compressive plastic strains (up to 25% thickness reductions) lead to an

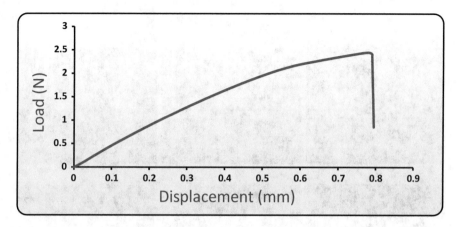

Fig. 4.33 Force versus displacement curve during testing of lap shear strength. The shear strength ($\sigma_s = F_{max}/A$), where F_{max} denotes the peak value of the force, and A is the overlap area for lap joint. This corresponds to a case in which a nominal plastic strain of 9.0% was imposed on E3/E15 in 1:1-42.3% PEG film, and a shear-strength of 0.07 MPa was noted

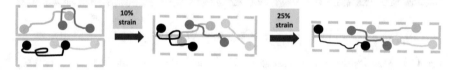

Fig. 4.34 Possible chain orientation in the stretch direction, which ultimately leads to reduced bonding

increased contact area (beyond discrete asperity contact), and, if factors other than chain interpenetration were responsible for bonding, one would expect a monotonic increase in G_c or σ_s with bonding plastic strain. The lowering of G_c or σ_s at large levels of plastic strains could be explained on the basis of anisotropic growth in microstructure, such that polymer chains preferentially re-orient in the direction of maximum principal stretch. We suggest that such chain orientation leads to a deviation from random-coil like configurations and, therefore, to less effective chain interpenetration across the interface, ultimately diminishing bonding at higher strains. This is illustrated schematically in Fig. 4.34. On the other hand, the fracture toughness of thermally-welded interfaces has been shown to increase monotonically as a function of time due to interdiffusion until the interface is fully healed [55].

A comparison of surface morphology before bonding and after fracture is shown in Fig. 4.35. The debonded surfaces indicate events of chain scissions or pullouts due to fracture; such features are similar to those reported on the fracture surfaces of polymers welded through interdiffusion [6, 16, 55] (also see Figs. 4.36 and 4.37).

To explicitly demonstrate the role of bulk plastic deformation in bonding (rather than merely a large contact traction), a "hydrostatic die" setup that prevents plastic deformation of circular stacked-up layers was designed. Compression experiments

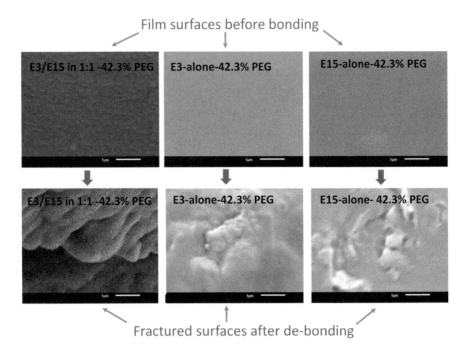

Fig. 4.35 SEM images of the top surfaces of E3/E15 in 1:1-42.3% PEG, E3-alone-42.3% PEG, and E15-alone-42.3% PEG, films before bonding and after debonding. The nominal plastic strains during roll-bonding for E3/E15 in 1:1-42.3% PEG, E3-alone-42.3% PEG, and E15-alone-42.3% PEG films were 15.53%, 8.12%, and 10.18%, respectively

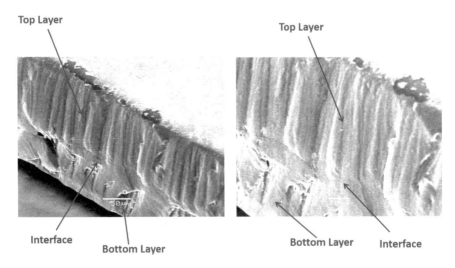

Fig. 4.36 SEM of the cross-section of a lap-joint prepared by upsetting, resulting in plasticity-induced-bonding

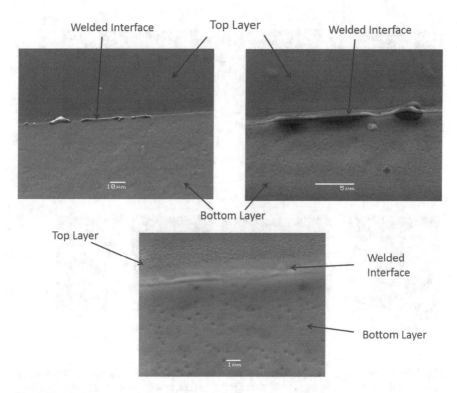

Fig. 4.37 SEM images on the lap joint prepared by plasticity-induced bonding

on identical film stacks, inside the "hydrostatic die" and using upsetting, were carried out in an Instron testing machine.

Figure 4.38 shows a comparison, where a stack of films (E3/E15 in 1:1-42.3% PEG), approximately 1″ in diameter, is compressed (i) without any constraints and, (ii) with the "hydrostatic die" constraint. The same level of peak compressive force (40 kN, resulting in a nominal compressive stress of 78.98 MPa) was applied at a loading rate of 1 mm/min in both cases. In the first case, the stack underwent macroscopic plastic flow, and layers bonded to form an integral structure. In the case of the "hydrostatic die" the layers splayed apart readily after removal. See [42] for detailed analyses.

An attempt to roll-bond high strength E3/E15 in 1:1-0% PEG film with E3/E15 in 1:1-42.3% PEG film was made, and a "no bonding" outcome, as shown in Fig. 4.39, was noted. E3/E15 in 1:1-42.3% PEG films demonstrated self-adhesion at much lower line loads (approximately 150 N/mm). As was shown in Fig. 4.20, 0% PEG films were relatively "hard" compared with 42.3% PEG films and, therefore, exhibited negligible plastic flow at the imposed load levels; 42.3% films were subjected to plasticity-induced molecular mobilization at lower loads. Importantly, activating bulk plastic flow on *both* sides of the interface, which leads

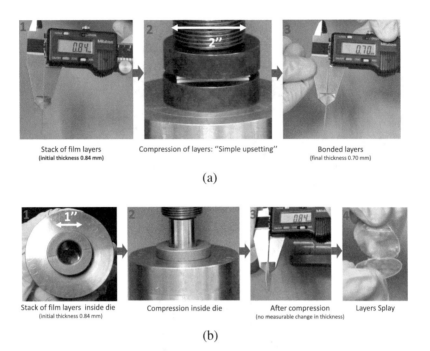

Fig. 4.38 Comparison of "simple upsetting" and "hydrostatic die" compression. (**a**) Compression of a stack of films without any die containment to permit macroscopic plastic flow and, thus, bonding. (**b**) Compression of a stack of films in a "hydrostatic die" that can generate large levels of hydrostatic pressure, but which limits plastic flow

to macromolecular interpenetration across the interface, is an essential requirement for plasticity-induced bonding.

The "no bonding" outcomes in the above two experiments also suggest that effects such as the possible presence of a rubbery layer at the surface, where the T_g on the surface may be lower than the bulk T_g and timescales for segmental relaxations may be relatively shorter, cannot by themselves lead to adhesion on the order of 1 s at approximately 60 K below the bulk T_g. The bonding below the bulk T_g (without any bulk plastic deformation), as reported in the literature, requires longer durations. Furthermore, existence of any enhanced relaxation of polymer chains (or segments) in the surface layer will be severely restricted by the glassy bulk beneath, so long-range diffusion within a short time is unlikely. We emphasize that the role of mechanical activation leading to enhanced mobility in the rubbery layer, during longer time healing experiments where moderate contact pressures were applied, has not been considered previously.

In principle, plastic deformation can cause a temperature rise due to irreversible mechanical work, and this could contribute to enhanced molecular mobility leading to bonding. The specific heat of E3/E15 in 1:1-42.3%PEG film was measured

Fig. 4.39 "No bonding" outcome between an "elastic" film and a "plastic" film: (**a**) 0% PEG film and 42.3% PEG film were fed into the rolling machine, (**b**) the rollers were brought closer and the loading was applied, (**c**) films exited from the rollers, and showed negligible adhesion. Line loading of approximately 250 N/mm was applied

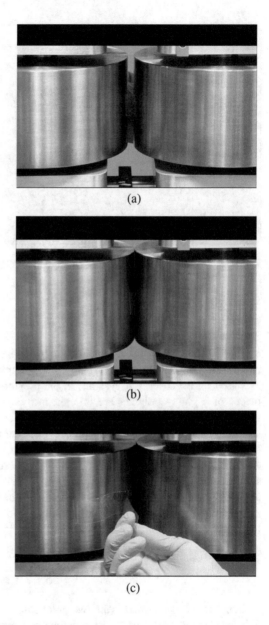

(a)

(b)

(c)

through differential scanning calorimetry (Fig. 4.40). The C_p was estimated to be 1860 J/kg-K, and the mass density was measured to be $\rho = 1180$ kg/m^3. Based on the stress strain curves, we estimated the flow stress for plastic deformation to be $\sigma_y = 4.12$ MPa. Accordingly, for a plastic strain as large as $\epsilon_p = 0.5$, the adiabatic temperature rise is estimated to be only:

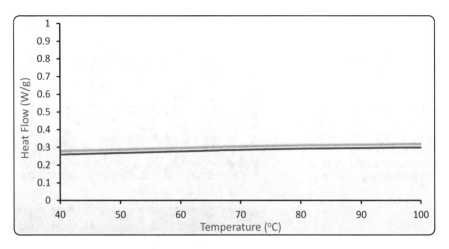

Fig. 4.40 DSC scan of E3/E15 in 1:1-42.3% PEG film. Calorimetry was done on a TA Q200

$$\triangle T = \frac{\sigma_y \epsilon_p}{\rho C_p} = 1.85 \,^{\circ}\text{C}.$$

As seen here, even fully adiabatic analysis gives only a negligible temperature rise considering that bonding is achieved approximately 60 K below the T$_g$.

Based on the several discussions in the previous chapters, it is clear that if two pieces of a glassy polymer are brought into contact within molecular proximity at temperatures well below their T$_g$ negligible adhesion, due to interdiffusion of macromolecules, will be noted. At well below the T$_g$ polymer chains are kinetically trapped [1, 21, 22, 47], time scales for relaxations in the glassy state are extremely large [15, 27, 28], and the system is frozen out with respect to cooperative segmental motions (α-like relaxation) [2] to exhibit interdiffusion. In [37], assuming a viscosity of 10^{13} Poise at T$_g$, self-diffusion coefficients of forty polymers were estimated to be around 10^{-25} m^2/s. The classic Bueche-Cashin-Debye equation [11, 12], which relates diffusivity (D) and viscosity (η), is given as:

$$\frac{D\eta}{\rho} = \left(\frac{A K_B T}{36} \right) \left(\frac{R^2}{M} \right). \tag{4.11}$$

In the above equation, A is the Avogadro constant, K_B is the Boltzmann constant, T is the absolute temperature, R^2 is the mean-square end-to-end distance of a single polymer chain, and M is the molecular weight. If we estimate D for our polymer at the T$_g$ (352 K) by considering $\eta = 10^{13}$ Poise, $\rho = 1180$ kg/m^3, $R^2 = 6 \times 7.4^2$ nm^2 (using R$_g$ of E3, and $R^2 = 6 \times$ R$_g^2$) and $M = 20, 300$ g/mol, $T = 352$ K, then $D = 1.12 \times 10^{-24}$ m^2/s. This is a remarkable estimate in terms of the order of magnitude, and compares well with the self-diffusivity of 10^{-25} m^2/s for several polymers at their

T_g, as estimated in [37]. Such small diffusivity near the T_g clearly indicates the aspect of kinetic arrest as the T_g is approached.

Consider a scenario in which a diffusion distance of $x = 10$ nm is to be achieved within a time (t) of 1 s; in such a case, $D(= x^2/2t)$ must be greater than 0.5×10^{-16} m^2/s. Based on the preceeding discussion, one can conclude that this is not possible, approximately 60 K below the bulk T_g. On the other hand, if one chooses a D value of 5×10^{-23} m^2/s at T_g then, to move a distance of 10 nm a time on the order of 10^8 s (approximately 3.16 years) would be required. This clearly shows that adhesion of glassy polymers, due to interdiffusion of macromolecules, near the T_g based on these diffusivity values cannot occur on very short experimental timescales. We remind readers that a glassy polymer may have a rubber-like layer on its free surface, which may exhibit faster timescales for relaxation processes; however, for reasons and results already discussed, the rubber-like layer can have a very limited role in a short time interval if the bulk of the polymer is several tens of degrees below T_g. Lastly, if the rubber-like layer was solely responsible for adhesion (without any aid from bulk plastic deformation) in time on the order 1 s, then it is not in accordance with the results of researchers who have reported adhesion of glassy polymers over longer timescales (up to several minutes, hours and days) on the grounds of a rubber-like layer.

We wish to again clarify the distinction of polymer welding above the T_g with respect to newly reported plasticity-induced molecular mobilization and bonding (which occurs within a period of time on the order of 1 s). Application of stresses leads to mechanically assisted enhanced mobility of polymer chains (or segments), and the polymer mobility in the respective cases is mechanistically quite different from each other. The exact micro-mechanisms or the mechanistic origins of relaxation processes at a molecular scale, and of interpenetration across the interface, are totally dependent on the molecular characteristics, and may be best-explored through detailed molecular-based computer simulations at this stage. In our opinion, however, this task remains an open question. The shear-transformation units of plastic deformation are also accompanied with local transient dilatations (i.e,. volume changes) that could allow several opportunities, such as dilatations providing "a hole" for interpenetration, to establish entanglements across the interface for bonding.

As a final comment, the self-diffusion coefficient (D) of a polymer chain in its melt state shows a strong dependence on molecular weight, i.e., $D \sim M^{-1}$ or $D \sim M^{-2}$) in accordance with the Rouse or the reptation model, respectively; however, all three blends of polymer considered here have been bonded in the order of 1 s, which is a significant contrast from the mechanism of polymer adhesion due to interdiffusion.

4.4   Note on Fracture Toughness of Polymer Interfaces Formed During Bonding

As mentioned earlier, large fracture energies of glassy polymers are due to the irreversible dissipation at the fracture tip. The toughness of glassy polymers comes from the events of chain disentanglement or scissions, which are viscous/plastic (irreversible) processes. The average interpenetration chain length and number of bridges formed across the interface play a major part in controlling the strength of the interface formed. In the case of a plastically welded interface, it is possible to have patches where chain interpenetration has not occurred (i.e., purely surface energy interactions act over those regions). However, the major contribution to toughness will be from regions where chain interpenetration has occurred and, during fracture, viscous drag or chain scission occurs. The overall G_{1c} will be given as a sum of the following components: frictional drag on chains during pull-out, chain-scissions, and surface energy interactions. The first two components dominate and can range up to several Joules/m^2, whereas surface interactions are at most a fraction of 1 J/m^2 [55]. Note, that the viscous-drag force on the chain is proportional to the separation rate so, at higher rates, toughness is expected to increase. At higher rates, scissions in polymer chains may also occur. High debonding stresses can also lead to crazing [16].

Debonding mechanisms change from simple chain-pullouts to the onset of crazing depending on whether weak or strong bonding has been achieved, respectively. Weak bonding will lead to brittle fracture during crack propagation and negligible plastic zone size ahead of the crack-tip. However, if strong bonding is noted, the debonding processes will involve large strain stretching and reorientation of chains in the pull-out directions, including cavitation or crazing, and cause a relatively large process zone in which inelastic failure occurs. Simulation of a large process zone with inelastic failure in molecular systems is not possible at the moment. Hence, the choice of cohesive zone models on the continuum scale is an appropriate choice for capturing key qualitative features from the molecular scale to the continuum scale model, along with the aid of macroscopic fracture experiments (or observations of experimental crack-tip processes).

4.5   Summary

Welding techniques are used widely for the joining or bonding of plastics. These include application of direct external heat, generation of heat through mechanical motion, or use of electromagnetism. The quality and choice of welding process directly dictates the performance of the formed product during service. The underlying mechanism of polymer bonding upon heating has been attributed to interdiffusion of polymer chains across the interface and formation of macromolecular entanglements which remain intact upon cooling. We have reported a new phenomenon in polymer bonding. We have clearly elucidated the effect of plastic deformation/mechanical stressing to cause molecular mobilization and chain interdiffusion, and challenged the existing theories of polymer bonding in the sense that researchers have ignored the effect of deformation during interdiffusion bonding. We have showed that polymeric films, based on HPMC and PEG prepared by solvent casting, could be bonded through deformation. Quantitative levels of adhesion, and their correlations with plastic strain, the morphologies of fractured interfaces after plastic welding, hydrostatic-die experiments, no-bonding outcomes between "elastic" and "plastic" film, and other related demonstrations served as key verifiers for this new technique of plastic welding. If bonding occurs due to interpenetration of chains across the interface during plastic deformation, then debonding comprises molecular (viscoplastic) events such as chain-pull outs and chain-scissions. These were supported by SEM images of the fractured surfaces bonded due to plastic welding. The role of deformation on molecular mobility and bonding above or below the T_g (or T_m) has not been completely understood. Similarly, the effects of accelerated conformational transitions or chain slippages have not been investigated or characterized on interdiffusion or bonding. Atomistic or molecular simulations can be utilized to reveal the underlying mechanisms of this unanticipated behavior. To validate the molecular hypotheses, development of a continuum-level constitutive model of bonding between plastically welded interfaces, and experimental verification of deformation-induced bonding under different conditions (material types and processing conditions) are required. Successful modeling, simulation, and experimental work promises to lay a firm foundation to study more complex material systems and applications. Examples of such applications include: (i) enhancing adhesion between miscible (or immiscible) polymers through use of plasticizing agents that promote molecular mobility; (ii) tailoring of appropriate scale surface patterning to exploit the opportunities of chain entanglement during deformation by increasing the surface area and interaction during contact; (iii) improving the shear-mixing characteristics of polymer blends or composite materials during processing; engineering functional material interfaces that are amenable to enhanced molecular mobility under moderate deformation to facilitate bonding through chain interpenetration; (iv) enabling adhesion between polymers that are particularly immiscible due to Fickian diffusion by imposing deformation, or using an intermediate material layer that creates anchors to both interfaces; (v) providing better explanations for the physical behavior of manufacturing processes where these effects have gone unnoticed; developing new bonding applications that utilize this new bonding methodology. High-molecular-weight polymers could be bonded by plastic deformation if local dilatation is accompanied by shear transformation that can aid establishment of entanglements and, if successful, the bottleneck of long healing times for such polymers will be overcome.

References

1. A. Alegria, E. Guerrica-Echevarria, L. Goitiandia, I. Telleria, J. Colmenero. Alpha-relaxation in the glass transition range of amorphous polymers. 1. Temperature behavior across the glass transition. Macromolecules **28**(5), 1516–1527 (1995)
2. C.A. Angell, K.L. Ngai, G.B. McKenna, P.F. McMillan, S.W. Martin, Relaxation in glassforming liquids and amorphous solids. J. Appl. Phys. **88**(6), 3113–3157 (2000)
3. Y.M. Boiko, Interdiffusion of polymers with glassy bulk. Colloid Polymer Sci. **289**(17–18), 1847–1854 (2011)
4. Y.M. Boiko, On the formation of topological entanglements during the contact of glassy polymers. Colloid Polymer Sci. **290**(12), 1201–1206 (2012)
5. Y.M. Boiko, Is adhesion between amorphous polymers sensitive to the bulk glass transition? Colloid Polymer Sci. **291**(9), 2259–2262 (2013)
6. Y.M. Boiko, R.E. Prud'Homme. Strength development at the interface of amorphous polymers and their miscible blends, below the glass transition temperature. Macromolecules **31**(19), 6620–6626 (1998)
7. Bond energies. http://chemwiki.ucdavis.edu/Theoretical_Chemistry/Chemical_Bonding/ General_Principles/Bond_Energies
8. M. Brogly, A. Fahs, S. Bistac, Assessment of nanoadhesion and nanofriction properties of formulated cellulose-based biopolymers by AFM, in *Scanning Probe Microscopy in Nanoscience and Nanotechnology 2*, NanoScience and Technology, ed. by B. Bhushan (Springer, Berlin, Heidelberg, 2011), pp. 473–504
9. H. Brown, Mixed-mode effects on the toughness of polymer interfaces. J. Mater. Sci. **25**(6), 2791–2794 (1990)
10. H. Brown, A molecular interpretation of the toughness of glassy polymers. Macromolecules **24**(10), 2752–2756 (1991)
11. F. Bueche, Viscosity, self-diffusion, and allied effects in solid polymers. J. Chem. Phys. **20**(12), 1959–1964 (1952)
12. F. Bueche, W. Cashin, P. Debye, The measurement of self-diffusion in solid polymers. J. Chem. Phys. **20**(12), 1956–1958 (1952)
13. B.-R. Cho, J.L. Kardos, Consolidation and self-bonding in poly (ether ether ketone) (peek). J. Appl. Polym. Sci. **56**(11), 1435–1454 (1995)
14. K. Cho, H. Brown, D. Miller, Effect of a block copolymer on the adhesion between incompatible polymers. I. Symmetric tests. J. Polym. Sci. Part B: Polym. Phys. **28**(10), 1699–1718 (1990)
15. R.H. Colby, Dynamic scaling approach to glass formation. Phys. Rev. E **61**(2), 1783 (2000)
16. C. Creton, E.J. Kramer, H.R. Brown, C.-Y. Hui, Adhesion and fracture of interfaces between immiscible polymers: from the molecular to the continuum scale, in *Molecular Simulation Fracture Gel Theory* (Springer, New York, 2002), pp 53–136
17. P.-G. de Gennes, Sur la soudure des polyméres amorphes. C. R. A cad. Sci. Paris B **291**, 219–221 (1980)
18. P.-G. de Gennes, The formation of polymer/polymer junctions. Tribol. Ser. **7**, 355–367 (1981)
19. P.-G. de Gennes, *Simple Views on Condensed Matter*, vol. 8 (World Scientific, Singapore, 1998)
20. P.-G. de Gennes, Soft Interfaces: The 1994 Dirac Memorial Lecture (Cambridge University Press, Cambridge, 2005)
21. P.G. Debenedetti, F.H. Stillinger, Supercooled liquids and the glass transition. Nature **410**(6825), 259–267 (2001)
22. M. Ediger, C. Angell, S.R. Nagel, Supercooled liquids and glasses. J. Phys. Chem. **100**(31), 13200–13212 (1996)
23. T. Ge, G.S. Grest, M.O. Robbins, Tensile fracture of welded polymer interfaces: miscibility, entanglements, and crazing. Macromolecules **47**(19), 6982–6989 (2014)

24. T. Ge, F. Pierce, D. Perahia, G.S. Grest, M.O. Robbins, Molecular dynamics simulations of polymer welding: strength from interfacial entanglements. Phys. Rev. Lett. **110**(9), 098301 (2013)
25. T. Ge, M.O. Robbins, D. Perahia, G.S. Grest, Healing of polymer interfaces: interfacial dynamics, entanglements, and strength. Phys. Rev. E **90**(1), 012602 (2014)
26. http://storage.dow.com.edgesuite.net/pharmaandfood-dow-com/pharma/Pharma_ METHOCEL_Comparison_Table.pdf
27. J.M. Hutchinson, Physical aging of polymers. Progr. Polym. Sci. **20**(4), 703–760 (1995)
28. B. Jerome, J. Commandeur, Dynamics of glasses below the glass transition. Nature **386**(6625), 589–592 (1997)
29. K.L. Johnson, *Contact Mechanics* (Cambridge University Press, Cambridge, 1987)
30. K. Jud, H.H. Kaush, J.G.J. Williams, Mater. Sci. **16**, 204 (1981)
31. H.H. Kausch, M. Tirrell, Polymer interdiffusion. Annu. Rev. Mater. Sci. **19**, 341–377 (1989)
32. C. Keary, Characterization of METHOCEL cellulose ethers by aqueous SEC with multiple detectors. Carbohydr. Polym. **45**(3), 293–303 (2001)
33. A. Kinloch, *Adhesion and Adhesives: Science and Technology* (Springer Science & Business Media, Dordrecht, 2012)
34. J. Klein, Interdiffusion of polymers. Science **250**(4981), 640–646 (1990)
35. D.B. Kline, R.P. Wool, Polymer welding relations and investigated by a lap shear joint method. Polym. Eng. Sci. **13**(11), 52–57 (1972)
36. K. Kunz, M. Stamm, Initial stages of interdiffusion of pmma across an interface. Macromolecules **29**, 2458–2554 (1996)
37. L.-H. Lee, Adhesion of high polymers. I. Influence of diffusion, adsorption, and physical state on polymer adhesion. J. Polym. Sci. Part A-2 Polym. Phys. **5**(4), 751–760 (1967)
38. L.-H. Lee, *Fundamentals of Adhesion* (Springer, Cham, 1991)
39. Methocel cellulose ethers in aqueous systems for tablet coating. http://www.dow.com/scripts/ litorder.asp?filepath=/198-00755.pd
40. Methocel molecular weight viscosity relationship. http://dowwolff.custhelp.com/app/answers/ detail/a_id/1316
41. W. Oliver, G. Pharr, An improved technique for determining hardness and elastic-modulus using load and displacement sensing indentation experiments. J. Mater. Res. **7**(6), 1564–1583 (1992)
42. N. Padhye, Sub-Tg, solid-state, plasticity-induced bonding of polymeric films and continuous forming. PhD thesis, Massachusetts Institute of Technology, 2015
43. N. Padhye, D.M. Parks, A.H. Slocum, B.L. Trout, Enhancing the performance of the t-peel test for thin and flexible adhered laminates. Rev. Sci. Instr. **87**(8), 085111 (2016)
44. N. Padhye, D.M. Parks, B.L. Trout, A.H. Slocum, A new phenomenon: sub-tg, solid-state, plasticity-induced bonding in polymers. Sci. Rep. **7**, 46405 (2017)
45. S. Prager, M. Tirrell, The healing process at polymer–polymer interfaces. J. Chem. Phys. **75**(10), 5194–5198 (1981)
46. S. Roy, C. Yue, Z. Wang, L. Anand, Thermal bonding of microfluidic devices: factors that affect interfacial strength of similar and dissimilar cyclic olefin copolymers. Sens. Actuat. B Chem. **161**(1), 1067–1073 (2012)
47. F.H. Stillinger, A topographic view of supercooled liquids and glass formation. Science **267**(5206), 1935–1939 (1995)
48. S.S. Voyutskii, Autohes. Adhes. High Polym. (Interscience, New York, 1963)
49. J. Washiyama, E.J. Kramer, C.F. Creton, C.-Y. Hui, Chain pullout fracture of polymer interfaces. Macromolecules **27**(8), 2019–2024 (1994)
50. K. Welp, R. Wool, G. Agrawal, S. Satija, S. Pispas, J. Mays, Direct observation of polymer dynamics: mobility comparison between central and end section chain segments. Macromolecules **32**(15), 5127–5138 (1999)
51. J.L. Willet, R.P. Wool, Macromolecules **26**, 5336 (1993)
52. R. Wool, K. O'connor, A theory crack healing in polymers. J. Appl. Phys. **52**(10), 5953–5963 (1981)

53. R. Wool, K. O'connor, A theory crack healing in polymers. J. Appl. Phys. **52**(10), 5953–5963 (1981)
54. R.P. Wool, Strength of polymer interfaces. Technical report, DTIC Document, 1990
55. R.P. Wool, *Polymer Interfaces: Surface and Strength* (Hanser Press, New York, 1995)

Chapter 5
Future Outlook

In an actively deforming glass, the origin of enhanced molecular mobility is associated with lowering of the surmounting potential barriers caused by the applied stresses. This enhanced molecular mobility significantly reduces the relaxation–time–spectrum of polymer chains or segments. The phenomenon of enhanced molecular mobility during deformation is not new, neither for polymers nor other material systems. However, it has seldom been used in any direct engineering application. This monograph has focused on presenting a unified view of polymer physics, continuum plasticity, and the origins of molecular mobility associated with the active deformation of glassy polymers. As of now, the most significant application of accelerated mobility in polymers during deformation is noted for polymer bonding. Results of the deformation-induced bonding of polymers have demonstrated that mechanical deformation can trigger the requisite polymer mobility for facilitating chain interpenetration and entanglements across the interface. The general concept underlying the deformation-induced polymer mobility for bonding promises new avenues for industrial applications and fundamental research.

Currently, the bonding between thermoplastics is achieved using adhesives, heat-based melting and solidification, surface modification, chemical reagents, or other treatments. Heat-based techniques are among the most commonly used methods, and are utilized in umpteen manufacturing processes, while relying on interdiffusion. Such methods include ultrasonic vibration, hot plates, radio frequency, implants, and laser or infrared. The application domains for heat-based methods span: packaging, lamination and sealing of components; automotive parts; medical devices; electrical appliances; biomaterials; fabrics; cosmetics; pharmaceuticals; space-applications. Some common examples of these applications are: transdermal patches [4] (formed from multiple layers of polymer films), microfluidic devices [3] (used in diagnoses of cancer cells and environmental sensing), plastic mulch films [6], greenhouse films, and packaging bags (manufactured by co-extrusion processes [16]), fabrication of actuators [19] for film-based soft-robotics, and nonwoven materials [11]. Last but not least, 3D printing methods that require melting

© The Author(s), under exclusive license to Springer Nature Switzerland AG 2021   95
N. Padhye, *Molecular Mobility in Deforming Polymer Glasses*,
SpringerBriefs in Materials, https://doi.org/10.1007/978-3-030-82559-1_5

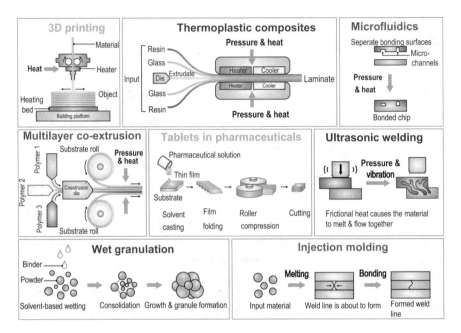

Fig. 5.1   Some industrial applications which rely upon polymer adhesion through interdiffusion

(or softening) and solidification of feed plastic material [5, 17], to form a desired part, rely on polymer bonding through interdiffusion. Figure 5.1 shows a few manufacturing applications where bonding between polymers is routinely achieved through interdiffusion.

The global market for laser plastic welding is on a course to reach a market value of US\$1.5 billion by 2025 [14]. The adhesive industry is estimated to be US\$15.0 billion in 2020 in the US alone [7]. In both these (and many other) sectors, polymer bonding through interdiffusion has a role of paramount importance. The main advantage of bonding polymers through welding (or interdiffusion) is that it can provide a molecular-level seal between two parts. On the other hand, joining of plastics via mechanical bolts does not achieve molecular-level connectivity, and the use of adhesives for bonding can lead to a "kissing joint" [8]. "Kissing joint" is a bonding defect in which the adhesive and the substrate are only in contact though a weak bond, and its occurrence is considered to be a result of contamination on the surfaces or due to environmental effects. "Kissing joints" are considered fatal and can act as a site for onset of premature failure of the interface. To avoid "Kissing joints", one could improve the bonding between plastic components by tailoring adhesives such that the adhesive placed at the interface between two plastics causes local plasticization and enhances molecular mobility and interpenetration into the respective sides upon mild compression (an approach proposed in [12] based on deformation-induced bonding).

Plastic joining through welding (or other heat-based) techniques is expected to have an increasingly important role because many industries are moving towards lightweight, flexible, resilient, and strong components made out of polymers [7]. Automakers and aircraft manufacturers have long recognized that weight reduction through use of plastics can make cars or aircrafts more energy efficient. Such lightweighting of parts, ranging from interior cabins, door assemblies, housings, fuel tanks, reservoirs, steering, and brakes, requires replacing conventional metals with polymers and hybrid or functional materials, and inevitably involves bonding between plastics. The efficacy of product assemblies in service depends upon seamless integration of its components, so achieving good polymer bonding is crucial.

The filler materials used in heat-based welding methods could be engineered in a manner (by tuning their physical and chemical properties) such that they can favorably undergo deformation-induced mobility and chain interpenetration across the interfaces. By utilizing the concept of deformation-induced mobility, it is expected that reductions in energy consumption can be achieved while minimizing direct heating. Such an approach will promote ecofriendly manufacturing practices, while optimizing manufacturing processes as a whole and product performance. The new deformation-induced bonding phenomenon is also expected to be applicable to existing and new genre of recyclable and biodegradable plastics; thus, promoting a circular and environmentally benign economy. However, to do so, one needs to understand and identify the mechanistic origins and attributes of deformation-induced molecular mobilization in polymers that cause bonding above and below T_g (or T_m in case of semi-crystalline polymers).

The newly proposed plastic welding discussed in this monograph is mechanistically quite distinct from the conventional thermal welding of polymers. In conventional diffusion bonding, without deformation, bonding is achieved through heating above the T_g (or T_m in case of semi-crystalline polymers) over the timescales of minutes to hours, with diffusion coefficient (D) in the melt state scaling as $D \sim M^{-1}$ or $D \sim M^{-2}$ (M being the molecular mass of the linear polymer), depending on the Rouse or reptation model of diffusion, respectively. Thus, for high-molecular weight systems, long healing times are required and practically no diffusion is noted at temperatures well below the T_g. In sub-T_g, solid-state, plasticity-induced bonding, polymer blends of different molecular weights have been bonded, well below the T_g, in the time on the order of a fraction 1 s, thereby alleviating the limitation of slowing down of diffusional dynamics owing to increasing molecular weight. These unprecedented experimental results have opened additional pathways of tailoring the bonding characteristics of polymer interfaces through a combination of temperature, time, and deformation.

Some challenges involved in developing a mechanistic understanding for the phenomenon of sub-T_g, solid-state, plasticity-induced bonding merit discussions. First, we do not exactly know the initial chain-end distribution of polymer chains on the free surface of polymeric materials (which depends on the method of preparation and processing conditions). Neither are we certain about their chemical and physical states at the free surface. For example, on the free surface there could exist weakly adsorbed ambient (or other) species,[1]a rubbery layer, or concentration gradients of additive molecules in the polymer matrix, which in general cannot be quantified deterministically at the molecular level during physical experiments. It is likely that the plastic relaxations of chain-ends/segments occurring once interfaces are in molecular proximity and subjected to compressive plastic deformation, are different from those happening in the bulk. This should be expected, because the thermodynamic state of chains (and their segments or ends) at the free surface is different from those in the bulk. The state of the free surface of a polymer could have a critical role in dictating the distribution, density, and orientation of chain ends. These factors can affect the bonding behavior under deformation because chain ends at the free surface are likely to aid interpenetration and entanglement during deformation. Statistical ensembles of thermodynamically consistent relaxation pathways that could lead to macroscopic observable behaviors on bonding should be regarded as the mechanistic origin of deformation-induced bonding (though it may be quite difficult to describe these processes precisely at this stage). Also, the operative mechanisms, of the polymer mobility at the interface during bonding, may be different from those reported in existing molecular studies for bulk plastic deformation.

Apart from the aforementioned challenges and uncertainties involved, the task of employing molecular simulations, to discover exact micro-mechanisms for an arbitrary polymer system during plastic welding at the interface poses some more challenges. Atomistic/MD simulations for polymer deformation can be seen in terms of following three steps: (1) assume model potentials, (2) validate the models on some known cases, and (3) run simulations or deformations to infer microscopic/molecular mechanisms from simulations. Clearly, choosing correct pair of inter-atomic potentials, that resembles the polymer/system at hand, is the key, which is challenging. At very fast deformation rates, usually done in molecular simulations for polymer deformation, the plastic deformation may exhibit a relaxation mechanism on a molecular level that may be different from the true mechanisms of plastic relaxations leading to bonding at experimentally reported (moderate) strain rates. The free-energy profile and thermodynamically feasible mechanisms operative at deformation rates, particularly at which bonding is noted experimentally, must be considered, and the plausible mechanisms of plastic relaxations inferred from high strain-rate simulations can be questioned for validity at moderate (or slower) strain rates (occurring in the physical experiments). For example, at slow strain rates very large cooperative rearrangements are likely; whereas, at very fast strain rates only segmental and or other levels of relaxations could occur. The reported deformation-induced bonding processes are observed in the time on the order of 1 s, and atomistic/molecular computer simulations are currently available for short time intervals (on the order of microseconds

---

[1] Adsorption can be of two types: (a) physio-sorption (e.g., gas molecules are held on solid surfaces due to van der Waals forces). (b) chemi-sorption (e.g., gas molecules are held on a surface through chemical bonds.

to nanoseconds). Thus, these are two very different regimes of operation. The key challenge with molecular simulations to reveal the mechanisms of plasticity-induced bonding for specific polymers lies in capturing the physical and chemical specificity of a polymer by representing its characteristics, such as chain lengths, bulky-side groups, steric effects, intramolecular-interacting forces, or chain rigidity. The microscopic mechanisms of plastic relaxations during the deformation are likely to depend on all of these factors. For relatively simple molecules, such as $(-(CH_2)_n-)$, this can be possibly done but, as a polymer chain becomes more complex, the task is quite challenging. We have speculated that the transient local dilatation accompanying shear-transformations could be a likely mechanism to facilitate enhanced inter-penetration and bonding. Although some molecular simulations of actively deforming polymers, below the T_g, do not show dependence of polymer mobility on free-volume fluctuations, these inferences must be extended with care while generalizing them to deformation at moderate rates, that too at the bonding interface. The free surfaces of real polymers can possess different scale of surface roughness, which could have a critical role in deformation-induced bonding. There is a possibility of achieving tailored bonding through purposely engineering the interface with topographical features. Although it is likely that stress concentration or plastic deformation will be concentrated at the location the surface-asperities, it is not clear whether the plastic deformation of nano-scale asperities alone will yield sufficient enhanced mobility to cause interpenetration in the absence of bulk activation. Experimental determination of such effects by analyzing free surfaces, prior to bonding or after debonding, is difficult through Scanning Electron Microscopy (SEM) because at a scale less than 10 nm the resolution of electron beams is likely to damage the free surface of the soft polymer.

The thermodynamic state or physical state at the interface is different from that in the bulk, and molecular simulations have their own limits. Hence, exact quantitative predictions for plasticity-induced mobility at the interface and chain-interpenetration are difficult under existing frameworks. The dynamic trajectories or molecular mechanisms associated with plastic deformation and bonding are open questions. Furthermore, a molecular-connectivity relationship that allows development of microscopic deformation modes for chain disentanglement during fracture of the plastically-welded interface is needed for phenomenological modeling of the failure of plastically welded interfaces. In plastic welding, the chain interpenetration is triggered during active plastic deformation of a polymer glass (non-equilibrium system), and is expected to be spatially heterogeneous, which adds to the complexity of modeling. This is unlike the situation in polymer melts, where growth of the interface can be captured reasonably well by diffusion models.

We wish to emphasize that the effects such as existence of a rubbery layer at the free surface of a polymer were not ruled out for sub-T_g, solid-state, plasticity-induced bonding; however, a rubbery layer with a glassy bulk beneath (several tens of degrees below the T_g), is argued to be incapable of demonstrating any long range diffusive motion in time on the order of second, by itself, to yield appreciable bonding. What is not certain is the exact role of a rubber-like layer, or whether the two pieces of a glassy polymer without the presence of a rubber-like layer on the free surface could be bonded through deformation? We note that active plastic deformation is associated with enhanced molecular mobility, but enhanced molecular mobility does not necessarily imply any appreciable chain interpenetration across interfaces, especially below the T_g (where molecular motion is effectively frozen). This scenario suggests that two pieces of a glassy polymer subjected to plastic deformation may not always necessarily show measurable adhesion due to polymer chain interpenetration. A similar argument is applicable to thermal welding of polymers around or below the T_g, and held under constant "pressure". Although, "non-hydrostatic" compression loadings lead to deviatoric stress components which can induce creep (and associated polymer mobility), this does not necessarily imply that interdiffusion occurs due to deviatoric stress. If mechanisms such as sub-Rouse relaxation or other lower-order chain/segments relaxations occur due to deviatoric stress instead of interdiffusion across the interface, then interpenetration cannot be attributed directly to mechanical activation.

Recently, we have obtained new results on the deformation-induced bonding of semi-crystalline polymers by adding plasticizers. Addition of plasticizers to semi-crystalline polymers led primarily to amorphous characteristics in the materials (according to X-ray diffraction). Addition of the plasticizer disrupted the crystalline domains and also promoted ductility. Semi-crystalline polymers without additives could hamper deformation-induced bonding in the solid state even if the amorphous domains are above their T_g because the crystalline domains may act to limit the molecular mobility of the chains from which they are comprised. By increasing the temperature, the mechanisms of plastic deformation in the crystalline domains of semi-crystalline polymers (transverse slip and chain slip [13]) might facilitate chain mobilization and bonding across the interface. Similarly, bonding of polymeric materials with foreign crystalline embeddings are expected to face difficulties in the absence of pronounced ductility (which is crucial for pharmaceutical manufacturing of oral-dose tablets from film-based, or alike, technologies [20]). Study of the mechanistic origins of chain mobility and interpenetration causing bonding in semi-crystalline materials poses similar (and perhaps greater) challenges in using atomistic/molecular simulations. One-to-one comparison and validation between experimental solid-state plasticity-induced bonding trends and those obtained through molecular simulations require correlation of the initial crystalline structures prior to bonding which, in turn depend upon the preparation history or crystallization kinetics that led to formation of the initial polymer systems. Although not explored yet, deformation-induced bonding of composite polymer materials (with nano- or micro-fillers) is another exciting venue of application that remains to be explored.

Direct means of experimentally probing in situ molecular-level motions during deformation-induced polymer mobility and chain interpenetration in *homopolymers* for rolling-like processes described earlier—through standard techniques such as neutron diffraction, X-ray diffraction, attenuated total reflection infrared (ATR IR), nuclear magnetic resonance (NMR), Raman spectroscopy, or confocal microscopy—are currently infeasible. X-ray and neutron diffraction have a small penetration depth. The amorphous polymer films that we used had a thickness of approximately 100 μm and a stack of several polymer layers forming a total thickness of 0.6–1.2 mm was subjected to roll-bonding, so obtaining information about precise molecular structures at a microstructurally disordered interface is not possible. ATR-IR, NMR, and Raman are not applicable since we are dealing with bonding of homopolymers, and there is no change in the chemical structure (or spectrum) across the interface after bonding. The depth of penetration is a limitation for applying these techniques. Confocal microscopy can provide a considerable depth of penetration, but fluorescent stains or probes mixed with the sample can themselves act as a plasticizer and alter the properties of the polymer matrix. Microscopy techniques such as SEM and TEM do not have a molecular-scale resolution. Attempts to study this phenomenon in roll-bonding would require accommodating a physically large, custom designed, roll-bonding machinery within an experimental analytical unit while triggering homogeneous and through-thickness plastic deformation necessary to cause bonding; this would be infeasible for in situ studies using existing analytical techniques. On the other hand, such limitations have not been encountered in studying the bonding of polymers through thermal welding. Therefore, diffraction techniques have been used in the past for estimating the diffusion coefficient and interfacial broadening. For example, Russell et al. [15] investigated the interdiffusion in a bilayer comprising a deuterated PS layer (320 Å thickness) on top of another protonated PS layer (1120 Å thickness) by neutron diffraction. In their experiments, reflectivity profiles showed good agreements with the expected sample thickness, and the change in scattering length density indicated that the deuterated macromolecules diffused into the protonated layer. Similar experiments were done by Kunz and Stamm [9] for PMMA films (90 nm thickness) to investigate the initial stages of interdiffusion above the T_g. By deuterating one of the components, they were able to study the broadening of the interface and the concentration profiles at a sub-nanometer resolution. Liu et al. [10] investigated the interdiffusion between PMMA films (100-nm thickness) above the T_g using X-ray diffraction and gold particles as markers. Alsten and Lustig [21] measured the diffusion coefficients in polymer melts of PS (thickness of 0.02–0.5 μm) and PMMA (thickness of 25–125 μm) by using ATR. Thus, several techniques have been used successfully for studying interdiffusion in thermal welding of polymers; however, their use for investigating plasticity-induced bonding is an open challenge.

Several manufacturing applications rely on molecular interdiffusion for bonding: (i) 3D printing; (ii) unit processes in pharmaceuticals (blending, mixing, granulation, compaction) in the presence of low-molecular-weight additives such as binders or those incurring local heating due to mechanical movement and collisions that cause contact-activated macromolecular diffusion; (iii) local plasticization of polymeric interfaces due to application of wet/solvent-based adhesives for enabling molecular diffusion and bonding; (iv) co-extrusion of multilayered hot extrudates in which different material layers form a bonded unit due to interdiffusion of macromolecules between miscible polymer interfaces under high temperatures and extrusion pressures [18]; (v) viscous calendaring of polymers through multistage rollers where mixing and rapid molecular interdiffusion can occur due to high temperatures and shearing action. In many of these processes, the explicit role of deformation-induced mobility to enhance polymer interdiffusion has been ignored completely or not reported. For example, in [22], authors explicitly recognized that bonding achieved in ultrasonic welding (in about 1 s) could not be explained based on the theory of reptation (which predicts several minutes for interdiffusion to occur). In ultrasonic welding of polymers, high interfacial temperatures are generated due to frictional heating, which causes local melting, and contact traction can provide mechanical activation to accelerate molecular mobility further and enhance bonding. Models for large-strain plastic deformation of bulk glassy polymers above and below the T_g are now widely available (e.g., see [1, 2]). Similarly, continuum-scale interfacial constitutive models of glassy (or semicrystalline) polymers undergoing bonding during active plastic deformation would be desirable. A multiscale approach—involving atomistic or molecular simulations, continuum-scale modeling, and experimentation—linking multiple length scales is a promising direction for reaping the full benefits of deformation-induced bonding. As discussed before, revealing exact molecular mechanisms poses great challenges, so one may need to be content with mechanistically motivated continuum-scale phenomenological models and experimental validations. Mechanistic insights are critical for gaining molecular-level understanding, and continuum-scale modeling is required to undertake fast simulations for predictions and validation in arbitrary 3D-deformation settings. Without atomistic simulations, the pathways to gain molecular insights are limited, whereas conducting atomistic simulations alone for making predictions in large-scale manufacturing processes pose computational impracticality. Thus, a multiscale approach utilizing molecular insights for developing continuum-scale simulations, where measurable properties can be used for simulations, is a logical and reliable route forward. Such an approach could capture the salient effects of molecular weights, polymer type, temperature, strain rate, contact mechanics, and lowering in the surface T_g of polymers on bonding due to deformation-induced mobility.

What happens exactly during deformation-induced bonding will, at best, stay an unresolved issue for now. However, what is needed to employ plasticity-induced mobility and bonding could be addressed more affirmatively through experimental and computational investigations, and is expected to find wide utility for practitioners.

References

1. L. Anand, M.E. Gurtin, A theory of amorphous solids undergoing large deformations, with application to polymeric glasses. Int. J. Solids Struct. **40**(6), 1465–1487 (2003)
2. M.C. Boyce, D.M. Parks, A.S. Argon, Large inelastic deformation of glassy polymers. Part I: rate dependent constitutive model. Mech. Mater. **7**(1), 15–33 (1988)
3. H.-J. Chiang, S.-L. Yeh, C.-C. Peng, W.-H. Liao, Y.-C. Tung, Polydimethylsiloxane-polycarbonate microfluidic devices for cell migration studies under perpendicular chemical and oxygen gradients. J. Visual. Exp. **2017**(120), e55292 (2017)

4. N.R. Council, *Polymer Science and Engineering: The Shifting Research Frontiers* (The National Academies Press, Washington, DC, 1994)
5. C.S. Davis, K.E. Hillgartner, S.H. Han, J.E. Seppala, Mechanical strength of welding zones produced by polymer extrusion additive manufacturing. Add. Manuf. **16**, 162–166 (2017)
6. E. Espi, A. Salmeron, A. Fontecha, Y. García, A. Real, Plastic films for agricultural applications. J. Plastic Film Sheet. **22**(2), 85–102 (2006)
7. IBISWorld, Adhesive manufacturing industry in the US - market research report (2020)
8. C. Jeenjitkaew, F. Guild, The analysis of kissing bonds in adhesive joints. Int. J. Adhes. Adhes. **75**, 101–107 (2017)
9. K. Kunz, M. Stamm, Initial stages of interdiffusion of pmma across an interface. Macromolecules **29**(7), 2548–2554 (1996)
10. Y. Liu, G. Reiter, K. Kunz, M. Stamm, Investigation of the interdiffusion between poly (methyl methacrylate) films by marker movement. Macromolecules **26**(8), 2134–2136 (1993)
11. S. Michielsen, B. Pourdeyhimi, P. Desai, Review of thermally point-bonded nonwovens: materials, processes, and properties. J. Appl. Polym. Sci. **99**(5), 2489–2496 (2006)
12. N. Padhye, D.M. Parks, B.L. Trout, A.H. Slocum, Plasticity induced bonding, 26 Feb. 2019. US Patent 10,213,960
13. D.M. Parks, S. Ahzi, Polycrystalline plastic deformation and texture evolution for crystals lacking five independent slip systems. J. Mech. Phys. Solids **38**(5), 701–724 (1990)
14. T.M. Research, Laser plastic welding market (2020). https://www.transparencymarketresearch.com/laser-plastic-welding-market.html
15. T. Russell, A. Karim, A. Mansour, G. Felcher, Specular reflectivity of neutrons by thin polymer films. Macromolecules **21**(6), 1890–1893 (1988)
16. W. Schrenk, T. Alfrey Jr., Coextruded multilayer polymer films and sheets, in *Polymer Blends* (Elsevier, Amsterdam, 1978), pp. 129–165
17. J.E. Seppala, K.D. Migler, Infrared thermography of welding zones produced by polymer extrusion additive manufacturing. Addit. Manuf. **12**, 71–76 (2016)
18. J. Song, A.M. Baker, C.W. Macosko, R.H. Ewoldt, Reactive coupling between immiscible polymer chains: acceleration by compressive flow. AIChE J. **59**(9), 3391–3402 (2013)
19. Y. Song, S. Qin, J. Gerringer, J.C. Grunlan, Unusually fast and large actuation from multilayer polyelectrolyte thin films. Soft Matter **15**(11), 2311–2314 (2019)
20. B.L. Trout, T.A. Hatton, E. Chang, J.M. Evans, S. Mascia, W. Kim, R.R. Slaughter, Y. Du, H.H. Dhamankar, K.M. Forward, G.C. Rutledge, M. Wang, A.S. Myerson, B.K. Brettmann, N. Padhye, J.-H. Chun, Layer processing for pharmaceuticals, 8 Dec. 2015. US Patent 9,205,089
21. J.G. Van Alsten, S.R. Lustig, Polymer mutual diffusion measurements using infrared ATR spectroscopy. Macromolecules **25**(19), 5069–5073 (1992)
22. R.J. Wise, *Thermal Welding of Polymers* (Woodhead Publishing, Sawston, 1999)

Printed in the United States
by Baker & Taylor Publisher Services